Thushara Weerawardane

Optimization and Performance Analysis
of High Speed Mobile Access Networks

T0215833

VIEWEG+TEUBNER RESEARCH

Advanced Studies Mobile Research Center Bremen

Herausgeber | Editors:

Prof. Dr. Otthein Herzog
Prof. Dr. Carmelita Görg
Prof. Dr.-Ing. Bernd Scholz-Reiter

Das Mobile Research Center Bremen (MRC) erforscht, entwickelt und erprobt in enger Zusammenarbeit mit der Wirtschaft mobile Informatik-, Informations- und Kommunikationstechnologien. Als Forschungs- und Transferinstitut des Landes Bremen vernetzt und koordiniert das MRC hochschulübergreifend eine Vielzahl von Arbeitsgruppen, die sich mit der Entwicklung und Anwendung mobiler Lösungen beschäftigen. Die Reihe „Advanced Studies" präsentiert ausgewählte hervorragende Arbeitsergebnisse aus der Forschungstätigkeit der Mitglieder des MRC.

In close collaboration with the industry, the Mobile Research Center Bremen (MRC) investigates, develops and tests mobile computing, information and communication technologies. This research association from the state of Bremen links together and coordinates a multiplicity of research teams from different universities and institutions, which are concerned with the development and application of mobile solutions. The series "Advanced Studies" presents a selection of outstanding results of MRC's research projects.

Thushara Weerawardane

Optimization and Performance Analysis of High Speed Mobile Access Networks

VIEWEG+TEUBNER RESEARCH

Bibliographic information published by the Deutsche Nationalbibliothek
The Deutsche Nationalbibliothek lists this publication in the Deutsche Nationalbibliografie;
detailed bibliographic data are available in the Internet at http://dnb.d-nb.de.

Dissertation University of Bremen, 2010

Gedruckt mit freundlicher Unterstützung des
MRC Mobile Research Center der Universität Bremen

Printed with friendly support of
MRC Mobile Research Center, Universität Bremen

1st Edition 2012

All rights reserved
© Vieweg+Teubner Verlag | Springer Fachmedien Wiesbaden GmbH 2012

Editorial Office: Ute Wrasmann | Anita Wilke

Vieweg+Teubner Verlag is a brand of Springer Fachmedien.
Springer Fachmedien is part of Springer Science+Business Media.
www.viewegteubner.de

Cover design: KünkelLopka Medienentwicklung, Heidelberg
Printed on acid-free paper

ISBN 978-3-8348-1709-9

Preface

This work was completed while employed as a research scientist at the Communication Networks Group (ComNets) of the Center for Computer Science and Information Technology (TZI), University of Bremen, Germany.

I would like to take this opportunity to sincerely thank Prof. Carmelita Görg. As the head of the Communication Networks Group, she provided valuable guidance, advice, and direction to complete my work successfully. I am deeply grateful to Prof. Andreas Timm-Giel, who supplied me with invaluable knowledge, encouragement, and supervision of my project and research at ComNets. I also thank Prof. Ranjit Perera, who gave me support, guidance, and advice on the analytical research portion of my work. Stephan Hauth, Dr. Gennaro Malafronte, and Thomas Reim deserve thanks for their profound advice and support in a number of technical matters. In addition, I am grateful to Dr. Andreas Könsgen, who not only provided valuable knowledge on scientific and technical matters, but also gave insight on other non-technical matters throughout my studies at the University of Bremen.

My colleagues in the project group, OPNET experts Yasir Zaki and Umar Toseef, deserve many thanks for their contributions as well as our many technical discussions. Asanga Udugama provided assistance and support with programming issues that I encountered during my research. Likewise, this work would not have been possible without my other colleagues: Dr. Koojana Kuladinithi, Markus Becker, Dr. Bernd-Ludwig Wenning, Liang Zhao, Chen Yi, Mohammad Siddique, Amanpreet Singh, Chunlei An, Gulshanara Singh and Vo Que Son all extended their support to successfully finalize this work.

There are many people in my non-academic life who have helped me in different ways. First and foremost, it is my loving wife, Ureka, who shouldered much of the family responsibility during my preoccupation with research, as well as sharing all of life's ups and downs with me. I am humbled by the love and affection of my son, Deshabhi, who had to sacrifice his play-time for my studies on many occasions. Last but not least, I remember, with respect, my loving parents; they live far away from me now, but provided me the foundation and the encouragement to reach new heights in life.

Thushara Weerwardane

Abstract

Cost-effective end-to-end performance is one of the key objectives in flat mobile broadband networks. Even though the radio channel is an important, scarce resource, the rest of the mobile access network must respond according to time variant radio channel capacity by guaranteeing required QoS. Failing this, good overall performance will not be achieved effectively. Packet traffic inside high-speed mobile access networks is often bursty in nature, which causes a severe impact on end-to-end performance. Therefore, proper control of data flows over broadband mobile access networks is necessary to optimize overall performance. The focus of this work is to introduce new transport features for High Speed Packet Access (HSPA) and Long Term Evolution (LTE) networks that provide better QoS guarantees and enhance overall end-to-end performance.

Novel adaptive flow control and enhanced congestion control algorithms are proposed, implemented, tested and validated using a comprehensive HSPA system simulator. During HSPA and LTE simulator development, innovative scheduling approaches are introduced and developed for both downlink and uplink streams. Both these system simulators provide great flexibility and enhanced scalability for the analysis of overall network performance, including all protocol layers from the application to physical layers. The impact of adaptive flow control and congestion control algorithms on end-to-end performance are investigated and analyzed under different overloading conditions in the backhaul network. These analyses confirm that the proposed algorithms not only enhance end user performance by providing guaranteed QoS, but also optimize overall network utilization. Thus, the algorithms are able to cost-effectively provide reliable and guaranteed services for both network operators and end users.

Two new analytical models, one for congestion control and the other for combined flow control and congestion control, both based on Markov chains, have been designed and developed to overcome simulation issues such as long development times and lengthy simulations. The proposed analytical models provide exceptional efficiency regarding speed of analysis as well as high accuracy compared to the detailed HSPA simulator. Therefore, the new analytical models can be used to evaluate performance of adaptive flow control and enhanced congestion control algorithms more effectively within a shorter period of time compared to simulation based analysis.

Contents

List of Figures

List of Tables

List of Abbreviation

2G	The Second Generation Mobile Communication Systems
3G	The Third Generation Mobile Communication Systems
3GPP	3rd Generation Partnership Project
AAL2	ATM Adaptation Layer 2
AAL5	ATM Adaptation Layer 5
ALCAP	Access Link Control Application Part
ATM	Asynchronous Transfer Mode
AUC	Authentication Center
BCH	Broadcast Channel
BER	Bit Error Rate
BS	Base Station
CAC	Call Admission Control
DCH	Dedicated Channel
DL	Downlink
DSCH	Downlink Shared Channel
FACH	Forward Access Channel
FDD	Frequency Division Duplex
FDMA	Frequency Division Multiple Access
FP	Frame Protocols
ETSI	European Telecommunication Standards Institute
GGSN	Gateway GPRS Support Node
GSM	Global System for Mobile Communications
HLR	Home Location Register
ID	Identifier
IP	Internet Protocol
ISDN	Integrated Services Digital Network
ISO	International Standards Organization
ITU	International Telecommunication Union
ITU-T	Telecommunication Standardization Sector of ITU
MAC	Medium Access Control
ME	Mobile Equipment
MM	Mobility Management
MSC	Mobile Services Switching Center
NBAP	Node-B Application Part
NSS	Network and Switching Subsystem
PCH	Paging Channel

PDU   Protocol Data Unit
PSTN   Public Switched Telephone Network
QoS   Quality of Service
QPSK   Quadrature Phase Shift Keying
RAB   Radio Access Bearer
RACH   Random Access Channel
RAN   Radio Access Network
RANAP   Radio Access Network Application Part
RLC   Radio Link Control
RNC   Radio Network Controller
RNS   Radio Network Subsystem
RNSAP   Radio Network Subsystem Application Part
RRC   Radio Resource Control
RTT   Round Trip Time
SIR   Signal-to-Interference Ratio
TCP   Transmission Control Protocol
TDD   Time Division Duplex
TDMA   Time Division Multiple Access
UDP   User Datagram Protocol
UE   User Equipment
UL   Uplink
UMTS   Universal Mobile Telecommunications System
UNI   User Network Interface
USIM   UMTS Subscriber Identity Module
UTRAN   UMTS Terrestrial Radio Access Network
VLR   Visitor Location Register
VoIP   Voice over Internet Protocol
WCDMA   Wideband Code Division Multiple Access

List of Symbols

$\Pr(y)$: Probability of y

λ_S: source rate

λ_D: drain rate

$PBR(t)$: provided bit rate, average number of MAC PDUs at time t

$\overline{PBR(t)}$: weighted average number of MAC PDUs at time t

qs: queue size

$f(t)$: buffer filling level

$\Delta(t_i)$: delay variation between t_i and t_{i+1}

$R(t_i)$: build-up delay variation between t_i and t_{i+1}

$A_n(t)$: congestion indication interarrival time

t_n: interarrival time between n^{th} and $(n+1)^{th}$ CI signals

X_n: state at the nth CI arrival

$p_{i,j}$: transition probability from state i to state j

β: reduction factor

$\alpha = 1 - \beta$

$\Pr(k)$: probability of exactly k CI arrivals within a step

π_n: n^{th} stationary state probability

P: transition probability matrix

$E(X) = $ expected mean value of random variable X or average value of X

$\overline{Y}$: expected mean value of random variable Y

r_k: k^{th} drop probability

V: Variance

pbr_j: j^{th} FC state probability

P_{fctofc}: transition probabilities for state changes within a FC session

q_n: probability of n CI arrivals within a given FC interval

q_0: probability of no CI arrivals within a given FC interval

n_{fc}: number of FC states

n_{cc}: number of CC states

n_{st}: number of time steps (FC time steps) within a CC state

n_t: total number of states

1 Introduction

High Speed Packet Access (HSPA) is introduced within the broadband wireless network paradigm as an extension of the Universal Mobile Telecommunication System (UMTS) which is standardized by the 3rd Generation Partnership Project (3GPP). The main objective of this technology is to enhance the data rate in the up- and downlink, and also to reduce the latency in both directions. In recent years, usage of real time and multimedia applications is rapidly increasing worldwide by demanding higher capacity and lower latency. In order to fulfill such requirements, 3GPP steps in by introducing the Long-Term Evolution (LTE) along with the System Architecture Evolution (SAE) as the foreseen broadband wireless access technologies.

Currently broadband wireless technologies are becoming a part of people's life style and a key requirement for every business worldwide. Therefore, high reliability of various services with different Quality of Service (QoS) requirements is vitally important. To fulfill such requirements, the complete system from end user to end user needs to be controlled in a cost-effective manner. If any part of the network does not comply with the rest, the overall performance degrades by wasting a large part of the valuable resources. From the end user perspective, the performance which can be measured in terms of user data throughput and QoS is the final outcome. To achieve such objectives, all parts of the networks should be properly dimensioned and controlled. The radio part of the broadband network which is widely in the research focus [17 and 18] is the main bottleneck of such achievements. However, in order to utilize the scarce radio resources efficiently, the rest of the network protocols should be adapted accordingly. The achievable capacity of the radio resources is time variant and also dependent on several other real time issues such as the traffic types and their QoS requirements, number of users and mobility, the environmental conditions etc. Such time varying radio capacity fluctuations have a huge impact on the rest of the network which reduces the cost-effectiveness and the overall performance [42, 43, and 44]. Since such issues regarding the overall performance are rarely addressed within the literature [24 and 25], one part of this work is focused on this area of investigations.

The proper dimensioning of the transport network is one of the key areas in mobile access networks to gain the aforementioned benefits for the end users. All transport and network protocols have to be parameterized in a suitable way in which they operate efficiently and cost-effectively, coexisting with the broadband mobile access network. For example, a limited transport network can cause congestion due to the unpredictable bursty nature of the traffic [3]. In such a situation, there is a requirement of an adaptive feedback flow control algorithm [4 and 7] which can closely monitor the time varying wireless capacity and control the input traffic to the transport network. Since a cost-efficient operation is vitally important for the mobile network operators (MNOs), often the transport network is dimensioned based on average network utilization [32 and 33]. Therefore congestion can occur during peak demands which are highly unpredictable in real scenarios. Depending on the severity, congestion can cause a significant impact on the overall performance and even obstruct the demanded services for a certain period of time. There are several protocols such as the Transmission Control Protocol (TCP) which are sensitive to these abrupt fluctuations [27, 29, and 30]. Since the customer satisfaction is one of the primary goals of the mobile network operators, such situations should be minimized or if possible avoided. It is identified that there is a clear requirement of the transport network flow control and congestion control for effective utilization of scarce radio resources to provide an optimum end user performance. Therefore, during the focus of this dissertation, issues related to the UTRAN network that can severely impact the end user performance and QoS experience are investigated and analyzed. As one of the main contributions of the author, this work introduces novel flow and congestion control algorithms for high speed packet access systems that can overcome all aforementioned issues and provide service guarantees to the end users while optimizing the overall network performance cost-effectively. The comprehensive detailed HSPA system simulation models have been developed by the author in the focus of the thesis and according to the requirements of the industrial research project managed by Nokia Siemens Network (NSN) to test, validate and analyze the above findings.

Investigations and analyzes using a detailed simulator along with the validation of the simulator itself are always time consuming activities which are sometimes unacceptably long. Therefore, often analytical approaches can provide faster investigation and analyzes along with a good accuracy compared to a simulation approach. For this reason, two analytical models are designed and developed by the author to evaluate the performance of the aforementioned flow control and congestion control algorithms in high speed broadband access networks.

Apart from the HSPA transport network, the work further extends the investigation and analyzes to the network and end user performance of the LTE transport network. Dimensioning the latter which completely operates on IP based packet networks for different QoS requirements is a key challenge for mobile network operators. For example, during LTE handovers, the traffic load over the S1 interface between the enhanced Node-B (eNB) and the Evolved

Packet Core (EPC) network and the X2 interface between two eNBs has to be efficiently controlled without degrading the end user QoS performance. Further, the forwarding data over the X2 interface has to be transferred without long delays in order to provide seamless mobility to the end users. The traffic prioritization over the transport network should be done carefully and effectively to meet the required QoS at the end users for different services. In such cases, the transport level congestion can worsen the impact on the end user performance by wasting overall network resources. Therefore, LTE transport network congestion should also be avoided by considering proper congestion control triggers. Further, an effective traffic differentiation model is deployed at the transport level in order to resolve the aforementioned issues. All these investigations and analyzes are performed by deploying suitable traffic models within the LTE system simulator which are designed and developed by the author. The effects of the LTE transport network (mainly the S1/X2 interface) during the intra-LTE handovers on the end-to-end performance are widely investigated and analyzed by introducing proper traffic differentiation models at the transport network level along with a comprehensive QoS aware MAC scheduling approach within the framework of this thesis.

This thesis work is organized as follows. Chapter 2 provides an overview of high speed broadband wireless networks such as UMTS, HSPA and LTE. First, the technological advancements of UMTS are discussed along with an architectural overview. Next, high speed access technologies are described by introducing the key enhancements in the down- and uplink. Several transport technologies such as ATM, IP and DSL which can be deployed for the high speed packet access network are also described in this chapter. New technological advancements of future broadband wireless networks are considered as well by introducing LTE along with an architecture overview. Further, all architectural changes to the previous technologies are presented by highlighting the prominent changes. The description of user mobility and LTE handovers are also described in this chapter.

Chapter 3 describes the design and development of a comprehensive HSPA simulator. The simulator development is performed in two steps: first the HSDPA simulation model is developed and then the HUSPA part is added. The chapter further discusses the challenges of developing a system simulator which is suitable for analyzing the transport and end-to-end performance. Apart from general protocol developments, novel MAC scheduling approaches for the downlink MAC-hs and the uplink E-DCH are introduced and implemented within the simulator whose details are given in this chapter. All UTRAN network entities along with the underlying transport technologies such as ATM and DSL are implemented according to the guideline and specification given by the 3GPP standards.

A new credit-based flow control algorithm is introduced in chapter 4. The technical and implementation aspects of the algorithm are described in detail. After defining appropriate traffic models and simulation scenarios, the performance of the algorithm is investigated and analyzed using the HSPA simulator. All achievements are given in the results analysis and the conclusion. In addition to the credit-based flow control algorithm a novel congestion control algorithm is also presented within the chapter. Different congestion detection principles are discussed along with implementation details. Variants of the congestion control algorithm are described by providing extensive investigations and analyzes using the simulator. The simulation results present the effectiveness of these novel approaches from the performance point of view. Finally the conclusion of the chapter summarizes the valuable findings about these new approaches.

Chapter 5 presents two novel analytical models which are based on the Markov property: first, a model for congestion control and second, a joint model for flow control and congestion control is developed. The theoretical backgrounds of both models are discussed in detail within the chapter. The outcome of these analytical models is compared with the simulation results. The chapter concludes by highlighting all achievements of the analytical models in comparison to the simulations.

Chapter 6 presents the conclusion of this work where all achievements are summarized and the outlook.

Appendix A presents effects of the DSL based UTRAN network on the performance of the HSPA network. The analysis is focused to investigate the impact of DSL packet losses, delay and delay variations – which are caused due to impairments of the DSL connections – on the HSPA network and the end user performance.

A description of the LTE network simulator is given in Appendix B. The main node models such as User Equipment (UE), enhanced Node-B (eNB) and the access Gateway (aGW) are designed and developed including peer-to-peer protocols such as TCP. The design of a proper IP DiffServ model and MAC scheduler is crucial for such simulator development since they have a great impact on the overall network performance. By considering all these challenges, the detailed implementation procedures of these network entities and the LTE handover modeling are presented in this chapter.

Appendix C presents an extensive analysis and investigation about the effects of the LTE transport network on the end user performance. For this analysis, a comprehensive LTE system simulator which includes a novel MAC scheduler and a novel traffic differentiation module are described in this chapter. The end user performance is evaluated for different overload situations at the transport network level by deploying appropriate traffic models. Further, the inter-eNB and intra-eNB user handover mobility is considered with different transport priorities for the forwarded traffic. The impact of all above transport effects on the different services at the end user is discussed in the results analyzes. The conclusion of the chapter summarizes all achievements of this investigation and analysis.

2   High Speed Broadband Mobile Networks

At the beginning of 1990, GSM with digital communication commenced as the 2nd generation mobile network and achieved staggering popularity of mobile cellular technology. With the evolution of wideband 3G UMTS, the usage of broadband wireless technology was further enhanced for many day-to-day applications. UMTS was primarily designed with dedicated channel allocation to support circuit-switched services however it was also designed with the motivation to provide better support for the IP based application than GPRS (General Packet Radio Service). Later, the 3GPP standard evolved into high-speed packet access technologies for downlink and uplink transmission [22]. They were popular in practice as HSDPA (High Speed Downlink Packet Access) and HSUPA (High Speed Uplink Packet Access). The latest step being investigated and developed in 3GPP is EPS (Evolved Packet System) which represents an evolution of 3G into an evolved radio access referred to as Long Term Evolution (LTE) and an evolved packet access core network in the System Architecture Evolution (SAE) [46]. Currently, the first deployment of LTE is entering the market [54].

As mentioned above, UMTS, HSPA and LTE are the main broadband wireless technologies in the 3rd and 4th generation mobile networks. The use of the Internet rapidly booms among the world population. In many cases, the Internet is accessed by handheld devices, using a large number of applications and services such as music downloads, online TV, and video conferencing. Most of these multimedia applications tend to demand for high speed broadband access [54]. On the other hand, today the basic platform for many activities such as business and marketing, daily routines and lifestyle of the people, medical activities and most of the social, cultural and religious activities are based on broadband wireless technology. In short, the technology becomes part of everyone's lifestyle.

The chapter is organized as follows. First, UMTS and related technologies are discussed along with the network architecture. Then new enhancements in the downlink and in the uplink are described under the heading of HSDPA and HSUPA respectively. Lastly the details about the LTE technology and its achievements are presented.

2.1   UMTS Broadband Technology

There is a huge change in the wireless paradigm after the introduction of the UMTS network. GSM is the first digital mobile radio standard developed for mobile voice communications. As an extension of GSM networks, GPRS has been introduced to provide packet switched data communications. UMTS triggers a phased approach towards an all-IP network by improving second generation (2G) GSM/GPRS networks based on Wideband Code Division Multiple Access (WCDMA) technology. UMTS supports backward compatibility with GSM/GPRS networks. GPRS is the convergence point between the 2G technologies and the packet-switched domain of the 3G UMTS [23].

Within this section, a brief overview of the UMTS technology, network architecture and its supported services are presented.

2.1.1   Wideband Code Division Multiple Access

The key enhancement of the radio technology from 2G mobile telecommunication networks to 3G mobile telecommunication networks (UMTS) is the introduction of WCDMA. It uses Direct Sequence (DS) CDMA channel access and Frequency Division Duplex (FDD) to access the radio channel among a number of users for the uplink and the downlink transmission. WCDMA transmits on a pair of radio channels with a bandwidth of 5 MHz each [17]. Further, WCDMA provides a significantly improved spectral efficiency and higher data rates compared to the 2G GSM network for both packet and circuit switched data. However, apart from these achievements, WCDMA also faces many challenges due to its complexity such as high computational effort at the receiver. Theoretically, it supports data rates up to 2 Mbit/sec in indoor or small cell out-door environments, and up to 384 kbps for the wide-area coverage [17 and 18].

2.1.2   UMTS Network Architecture

The UMTS architecture consists of the User Equipment (UE), the Universal Terrestrial Radio Access Network (UTRAN) and the Core Network (CN). The UMTS architectural overview is shown in Figure 2-1.

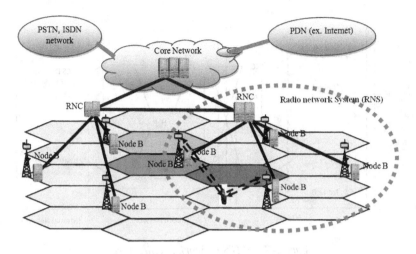

Figure 2-1: UMTS network overview

Figure 2-1 illustrates the hierarchical structure of UMTS starting from the Core Network to the User Equipment in the cell. The Core Network is the backbone of the UMTS network. The user equipments represent the lowest level of the hierarchy; UEs are connected to the Node-B (also called UMTS base station) via the radio channels. Several NodeBs are connected to a Radio Network Controller (RNC) through the transport network which includes a number of ATM based routers in between. The RNC is directly connected with the backbone network via the Iu interface. Further details about the UMTS network architecture are described by categorizing it into two main groups, the Core Network and the UTRAN.

2.1.2.1   *UMTS Core Network and Internet Access*

As shown in Figure 2-2, the core network provides the connection between the UMTS network and the external network such as the Public Switched Telephone Network (PSTN) and the Public Data Network (PDN). In order to perform a connection to the external networks with appropriate QoS requirements, the CN operates in two domains: circuit switched and packet switched. The circuit switched part of the CN network mainly consists of a Mobile service Switching Centre (MSC), a Visitor Location Register (VLR) and Gateway MSC (GMSC) network entities. All circuit switched calls including roaming, inter-system handovers and the routing functionalities are controlled by these main circuit switched network elements.

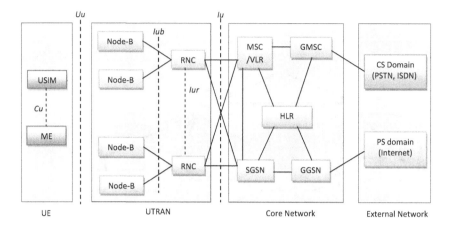

Figure 2-2: UMTS network architecture

In contrast to circuit switched domain activities, the packet switched domain mainly controls the data network. The packet switched part of the Core Network consists of the Serving GPRS Support Node (SGSN) and the Gateway GPRS support Node (GGSN). The SGSN performs the functions related to the packet relaying between the radio network and the Core Network along with the mobility and session management. The GGSN entity provides the gateway access control functionality to the outside world.

Apart from the network elements described above, there are several other network elements such as the Home Location Register (HLR), the Equipment Identity Register (EIR) and the Authentication Centre (AuC) in the Core Network. These network elements provide services to both packet-switched and circuit-switched domains. All subscribers' related information such as associated telephone numbers, supplementary services, security keys and access priorities are stored in the HLR. The EIR and the AuC are also acting as databases which keep identity information of the mobile equipment and security information such as authentication keys respectively.

2.1.2.2   UMTS Radio Access Network

The UTRAN consists of one or more radio network subsystems (RNS) and is shown in Figure 2-2. A radio network subsystem includes a radio network controller (RNC), several Node-Bs (or UMTS base stations) and many UEs. The radio network controller is responsible for the control of radio resources of UTRAN [44] and plays a very important role in power control (PC), handover control (HC), admission control (AC), load control (LC) and packet scheduling

(PS) algorithms. The RNC has three interfaces: the Iu interface which connects to the core network, the Iub interface which connects to Node-B entities and the Iur interface which connects with peer RNC entities.

The Node-B is comparable to the GSM base station (BS/BTS), and it is the physical unit for radio transmission and reception within the cells. The Node-B performs the air interface functions which include mainly channel coding, interleaving, rate adaptation, spreading, modulation and transmission of data. The interface which provides the connection with the user equipment is called the Uu interface. This radio interface is based on the WCDMA technology [56]. The Node-B is also responsible for softer handovers in which the UE is connected to a single Node-B with more than one simultaneous radio links whereas for the soft handovers the UE is connected with two Node-Bs with more than one simultaneous radio link [44]. During the process of softer handover, the Node-B is responsible of adding or removing the radio links which are connected with the UE that is located in the overlapping area of adjacent sectors of the same Node-B [44, 45]. The user equipment is based on the same principles as the GSM mobile station (MS), and it consists of two parts: mobile equipment (ME) and the UMTS subscriber identity module (USIM). Mobile equipment is the device that provides radio transmission, and the USIM is the smart card holding the user identity and personal information. The UTRAN user plane and the control plane [55] protocols are shown in Figure 2-3.

The user plane includes the Packet Data Convergence Protocol (PDCP), the Radio Link Control (RLC) protocol, the Medium Access Control (MAC) Protocol and the physical layer whereas the control plane includes all these protocols except the PDCP and the Radio Resource Control (RRC). The RRC is a signaling protocol which is used to set up and maintain the dedicated radio bearers between the UE and the RNC. The physical layer provides transport channels to the MAC layer at the interface. The MAC layer maps logical channels to transport channels and also performs multiplexing and scheduling functionalities. The RLC layer is responsible for error protection and recovery. The PDCP works in the user plane providing IP header compression and decompression functions [57].

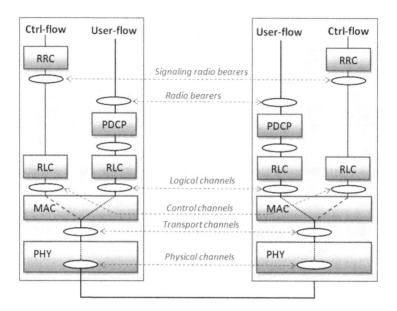

Figure 2-3: UTRAN control plane and user plane protocols

2.1.3   UMTS Quality of Service

UMTS has been designed to support a variety of quality of service (QoS) requirements that are set by end user applications. 3G services do vary from simple voice telephony to more complex data applications including voice over IP (VoIP), video conferencing over IP (VCoIP), video streaming, interactive gaming, web browsing, email and file transfer. The 3GPP has identified four different main traffic classes for UMTS networks according to the nature of traffic and services. They are termed conversational class, streaming class, interactive class and background class [17].

Real time conversation is always performed between peers of human end users. This is the only traffic type where the required characteristics are strictly imposed by human perception. Real time conversations are mainly characterized by the transfer time (or delay) and time variation (or jitter). These two QoS parameters should be kept within the human perception. The streaming class is mainly characterized by the preserved time variation (or jitter) between information entities of the stream, but it does not have tight requirements on low transfer delay as required by the voice applications. Therefore, the acceptable delay variation over transmission media is much higher than required by the applications of the conversational class. An example of this scheme is a user watching a real time video or listening to real time audio.

The interactive class of the UMTS QoS is applied when an end user (either human or a machine) is requesting data online from a remote entity such as a server or any other equipment. The basic requirements for this QoS class are characterized by a low RTD (Round Trip Delay), low response time and low bit error rate which preserves the payload contents. Web browsing, database retrieval and server data access are some examples for the interactive class. As the last category of this QoS classification, there is the background traffic class which includes applications such as email and Short Message Services (SMS) as well as download of files. The RTD and response time are not that critical for this QoS class but the data integrity must be preserved during the delivery of the data, therefore a low bit error rate is demanded for the transmission.

2.2 High Speed Packet Access

As discussed in the previous section, the UMTS implementation supports data rates up to 2 Mbit/s. However this is a theoretical limit; the practically feasible data rate is much smaller than the theoretical maximum. Therefore UMTS is suitable for most of the normal voice based applications and some of the Internet based applications such as simple web browsing and large file downloads. The usage of current Internet based applications is vastly growing worldwide among all age segments of people [20]. Traffic over any network becomes more bursty and demanding high data rates. Applications such as high quality video downloads, online TV and video conferencing require higher data rates and also lower delays. Furthermore, the quality perceived by the end user for interactive applications is largely determined by the latency (or delay) of the system. In order to satisfy these ongoing and future demands, the WCDMA air interface is improved in the 3GPP Rel'5 and Rel'6 specifications by introducing high speed downlink packet access (HSDPA) and high speed uplink packet access (HSUPA) respectively. HSUPA is officially referred to as E-DCH (Enhanced Dedicated Channel) in the 3GPP specification but industry widely uses the term HSUPA as a counterpart for HSDPA. To achieve a very high data rate and low latency both in the uplink and the downlink, WCDMA introduces three main fundamental technologies: fast link adaptation using Adaptive Modulation and Coding (AMC), fast Hybrid ARQ (HARQ) and fast scheduling [22]. These rely on the rapid adaptation of the transmission parameters to the instantaneous radio channel conditions in an effective manner in order to achieve higher spectral efficiency for the transmission. Before getting into the detailed discussion of the high speed downlink packet access (HSDPA) and high speed uplink packet access (HSUPA), a brief overview of these key enhancements of the above technologies is discussed in the following sections.

2.2.1   Adaptive Modulation and Coding

The basic principle of Adaptive Modulation and Coding (AMC) is to offer a link adaptation method which can dynamically adapt the modulation scheme and coding scheme to current radio link conditions for each UE. The modulation and coding schemes can be selected to optimize user performance in the downlink and the uplink when the instantaneous channel conditions are known. The users who are close to the Node-B usually have a good radio link and are typically assigned to higher-order modulation schemes with higher code rates (example: 64 QAM with R=3/4 turbo codes). The modulation order and/or code rate will decrease as the distance of a user from the Node-B increases. Higher-order modulation such as 64-QAM provides higher spectral efficiency in terms of bit/s/Hz compared to QPSK or BPSK based transmissions. Such schemes can be used to provide instantaneous high peak data rates especially in the downlink, when the channel quality is sufficiently good with high signal-to-noise-ratio (S/N). WCDMA systems are typically interference limited and rely on the processing gain to be able to operate at a low signal-to-interference ratio. The use of a higher order modulation becomes impossible in practice in multi-user environments due to the fact that it has limited robustness against interference. However, in the downlink, a large part of the power is allocated to a single user at a time, and therefore a large signal-to-interference ratio can be experienced by allowing higher-order modulation to be advantageously used. Hence, higher-order modulation combined with fast link adaptation is able to adapt the instantaneous channel condition effectively mainly for the downlink transmissions rather than the uplink transmission.

2.2.2   Hybrid ARQ

The principle of Hybrid ARQ (HARQ) is to combine retransmission data with its previous transmissions which were not successful prior to the decoding process at the receiver. Such HARQ mechanisms greatly improve the performance and add robustness to link adaptation errors [23]. For the packet-data services, the receiver typically detects and requests a retransmission of erroneously received data units. Combining the soft information from both the original transmission and any subsequent retransmissions prior to decoding will reduce the number of required retransmissions. This results in a reduction of the delay and robustness against link adaptation errors. The link adaptation serves the task of selecting a good initial estimate of the amount of required redundancy in order to minimize the number of retransmissions needed, while maintaining a good system throughput.

The hybrid ARQ mechanism serves the purpose of fine-tuning the effective code rate and compensates for any errors in the channel quality estimates used by the link adaptation. However HARQ will also introduce redundancy into the system which causes a lower utilization of radio resources. Proper link adaptation together with HARQ mechanisms will help to achieve effective bandwidth utilization as a total.

2.2.3   Fast Scheduling

The scheduler which moves from the RNC to the Node-B for HSPA is an important element in the base station that can effectively allocate the radio resources for the users by considering various aspects such as QoS requirements, instantaneous channel quality etc. The design of a proper base station scheduler is a complex task due to the different user requirements and also mobile network operators. However, designing a fast scheduler in any system has to primarily consider the radio channel. It has to change the resource allocation depending on the varying channel conditions. Therefore a small scheduling interval such as 2 milliseconds is often chosen to keep this flexibility of channel adaptation. Another important aspect which should be considered when designing a fast scheduler is to achieve high data rates for each UE satisfying delay requirements. The overall throughput and the fairness become a trade-off situation for any scheduler design. When one feature is optimized, the other is reversely affected. For example if a scheduler is designed to optimize the throughput, it selects the user with best channel quality and allocates full resources to that particular UE. If the UE has sufficient data in the transmission buffer, it can use even total radio resources during the period of best channel condition. This type of scheduling approach is commonly named Maximum C/I (MaxC/I) or channel dependent scheduling in the literature [44 and 54]. The opposite of this approach is the fair scheduling approach which considers the fairness among the users and provides guaranteed delay for each connection [44]. Often current mobile operators consider both aspects of this trade-off and design a scheduler which provides high throughput while providing required QoS guarantees to users.

2.2.4   High Speed Downlink Packet Access

HSDPA is an extension for the UMTS network that provides fast access for the downlink by introducing advanced techniques which were described above. Therefore, from the architectural point of view it uses the same architecture as UMTS Rel'99. In addition to the techniques such as fast link adaptation, fast hybrid ARQ and fast scheduling, there are some functional changes in UE, Node-B and RNC for the downlink. HSDPA introduces a new type of transport channel called High Speed Downlink Shared Channel (HS-DSCH) [22]. The HS-DSCH transport channel is mapped onto one or more High Speed Physical Downlink Shared Channels (HS-PDSCHs) depending on the instantaneous data rate.

The HS-PDSCHs operate on 2 millisecond transmission time intervals (TTIs) or sub-frame rather than the standard 10 milliseconds TTI which is used in UMTS Rel'99 and have a fixed spreading factor of sixteen [22]. Shorter TTIs have a better adaptation to the varying radio channel conditions, can achieve high interleaving gain and also provide much better delay performance. Further, this helps to provide better scheduling flexibility and granularity for HSDPA.

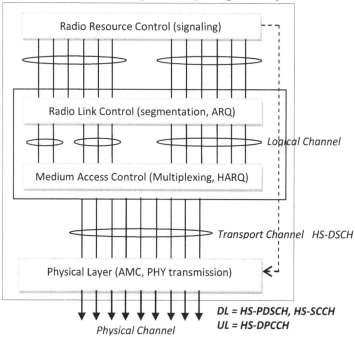

Figure 2-4: WCDMA channel and layer hierarchy for HSDPA

Figure 2-4 shows the WCDMA channels and layer hierarchy for an HSDPA system. Logical channels operate between Radio Link Control (RLC) and Medium Access Control (MAC). Transport channels are defined between MAC level and PHY level. Finally, physical channels are defined by the radio interface. Mobile stations in a cell share the same set of HS-PDSCHs, so a companion set of High Speed Shared Control Channels (HS-SCCHs) are used to indicate which mobile station should read which HS-PDSCH during a particular 2ms TTI. On call establishment of each mobile station a unique identifier called the H-RNTI (HS-DSCH Radio Network Transaction Identifier) and a set of HS-SCCHs are assigned. Whenever the network wishes to send some data to the mobile station it will setup the HS-SCCH using that mobile station's identity and also other network information which is required by the mobile station such as the number of HS-PDSCHs, their channelization codes and the HARQ process number.

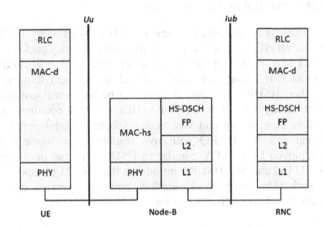

Figure 2-5: HSDPA UTRAN protocol architecture

Further, a new Node-B entity introduced at the MAC layer is called MAC-hs [59]. Most of the scheduling functions are shifted to this layer in the Node-B compared to standard UMTS. Further, the HARQ process runs at this level providing acknowledgement (ACK) and negative acknowledgements (NACK) for correctly and incorrectly received MAC PDUs respectively. Depending on whether the Transport Block (TB) was received correctly or not, the HARQ process of the mobile station will request its physical layer to transmit an ACK or NACK on the uplink HS-DPCCH channel for HSDPA DL transmission. There are between 1 and 8 HARQ processes running in parallel on any given HSDPA connection process.

The basic HSDPA UTRAN protocol architecture is shown in Figure 2-5 [22]. It shows both Uu and Iub interfaces and the respective protocols. In order to send the data over the Uu interface, data has to be delivered from RNC to Node-B appropriately. The UE demands for radio channel vary rapidly and therefore adequate data should be available at the MAC-hs buffers to cater such varying demands. Data packets which are sent by the MAC (MAC-d) layer in the RNC are called MAC-d PDUs. They are transmitted through the Iub interface as MAC-d flows or HS-DSCH data streams on the HS-DSCH transport channels. These HSDPA transport channels are controlled by the MAC-hs entity in the Node-B. Further, the HS-DSCH Frame Protocol (FP) mainly handles the data transport through the Iub interface, i.e. the interface between the RNC and the Node-B. All other protocols in the RNC still provide the same functionalities as used in UMTS. Ciphering of the data is one example of such functionality provided by the RNC that is still needed for HSDPA.

2.2.5   High Speed Uplink Packet Access

As discussed in the previous section, a new E-DCH is specified by 3GPP Rel'6 in order to provide an efficient mechanism for transferring bursty packet data traffic over the WCDMA uplink. It offers enhancements of the WCDMA uplink performance such as higher data rate, reduced latency and improved system capacity. Further, HSUPA has been designed to be backwards-compatible with existing functionality so that non-HSUPA UEs can still communicate with a HSUPA-capable base station and vice versa. Adaptive modulation and coding, fast scheduling and fast HARQ with soft combining are three fundamental techniques deployed by HSUPA. Similar to HSDPA, it also introduces a short uplink 2ms TTI [60 and 61]. HSUPA introduces the E-DCH transport channel to transfer the packets in the uplink. Figure 2-6 shows the overview of the channel and layer hierarchy for the HSUPA system.

As shown in Figure 2-6, several new physical channels are introduced for the uplink transmission. The Enhanced Dedicated Physical Data Channel (E-DPDCH) is defined to carry the user data bits to the network where the bits from this channel are delivered up to the MAC-e through the E-DCH transport channel. Furthermore, E-DPDCH uses a variable Spreading Factor (SF), for example the maximum data rate is achieved with two times SF-2 codes plus two times SF-4 codes [61]. All control information which is required by the base station to decode E-DPDCH data is sent via the Enhanced Dedicated Physical Control Channel (E-DPCCH). E-DPDCH and E-DPCCH are the main physical channels which are added in the uplink whereas three other new physical channels are defined in the downlink direction. They are the Enhanced Hybrid Indicator Channel (E-HICH) which carries the ACKs and NACKs to the UE, Enhanced Absolute Grant Channel (E-AGCH) which is a shared channel that signals absolute values for the Grant for each UE with a unique E-RNTI identity, Enhanced Relative Grant Channel (E-RGCH) which signals the incremental up/down/hold adjustments to the UE's Serving Grant. The E-RGCH and E-HICH share the same code space in the Orthogonal Variable Spreading Factor (OVSF) tree [61]. Orthogonality between the two channels is provided by the use of orthogonal 40 bit signatures which are the limited resources for the uplink transmission, because up to 40 different signatures can be encoded. In contrast to HSDPA, HSUPA does not utilize a shared channel for data transfer in the uplink. Each UE has a dedicated uplink connection which is realized by a unique scrambling code. In contrast to this, Node-B uses a single scrambling code and then assigns different OVSF channelization codes to differentiate UEs in the downlink [22].

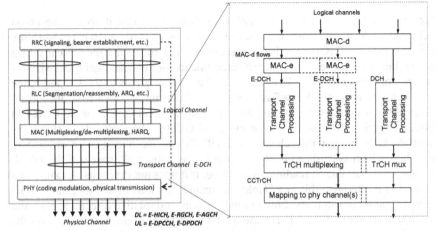

CTrCH -> Common Transport Channel, E-DCH -> Enhanced Dedicated Channel, CCTrCH = Coded Composite Transport Channel
E-HICH = E-DCH Hybrid ARQ Indicator Channel, E-RGCH = E-DCH Relative Grant Channel, E-AGCH = E-DCH Absolute Grant Channel
E-DPCCH = Enhanced Dedicated Physical Control Channel, E-DPDCH = Enhanced Dedicated Physical Data Channel

Figure 2-6: WCDMA channel and layer hierarchy for HSUPA

2.2.5.1   Uplink Shared Resource and Grants

The main shared resource in the uplink is the interference level in a cell. Managing the interference is done via a fast closed loop power control algorithm. The primary shared resource on the uplink is the total power received at the Node-B for a particular cell. Hence HSUPA scheduling is performed by directly controlling the maximum amount of power that a UE can use to transmit at any given point in time. Therefore, one of the primary goals of HSUPA is to achieve effective fast scheduling which allows adapting to rapidly changing radio channels with different data rates [18]. On the other hand, with 2 ms TTIs, the overall transmission delay is greatly reduced. The transmission delay performance is further improved by introducing HARQ technique as used in the HSDPA network.

HSUPA mainly uses two types of resource grants in order to control the UE's transmit power: non-scheduled grants and scheduled grants. The non-scheduled grant is most suited for constant-rate delay-sensitive applications such as voice-over-IP. In the non-scheduled grant which is mapped to a certain power level at the UE, the Node-B simply tells the UE the maximum Transport Block Size (TBS) that it can transmit on the E-DCH during the next TTI. The TBS is signaled at call setup and the UE can then transmit a transport block of that size or less in each TTI until the call ends or the Node-B modifies the non-scheduled grant via an RRC reconfiguration procedure.

For scheduled grants, the UE maintains a serving grant that it updates based on information received from the Node-B via E-AGCH or E-RGCH downlink channels [61]. E-AGCH signals the absolute serving grants and the UE can adjust its maximum power level in order to determine maximum transport block size for the current transmission. E-RGCH signals the relative grants to the UE and based on this information the UE adjusts its serving grant up or down from its current value. At any given point in time the UE will be listening to a single E-AGCH from its serving cell and to one or more E-RGCHs. The E-RGCH is shared by multiple UEs but on this channel the UE listens for a particular orthogonal signature which consists of a 40 bit code in the same code space of the OVSF (Orthogonal Variable Spreading Factor) tree. If it does not detect its signature in a given TTI it interprets this as a "Hold" command, and thus makes no change to its serving grants. In summary, both grants, absolute and relative directly specify the maximum power that the UE can use on the E-DPDCH in the current TTI. As the E-DCH block sizes map deterministically to power levels [18], the UE can translate its Serving Grant to the maximum E-DCH transport block size which can be used in a TTI.

2.2.5.2   *Serving Radio Link Set and Non-Serving Radio Link Set*

The concept of a Serving Radio Link Set (S-RLS) and a Non-Serving Radio Link Set (Non-SRLS) is defined in combination with soft handover for HSUPA [18, 43]. The group of cells from which the UE can soft combine E-RGCH commands, create the serving RLS. The serving RLS by definition includes the serving cell from which the UE is receiving the E-AGCH. Further, the cells in the serving RLS must all transmit the same E-RGCH command in each TTI, which means the cells that are belonging to the same RLS should be controlled by the same Node-B. Apart from this, the UE can also receive the E-RGCH information of any other cell which belongs to another RLS. All such cells that transmit an E-RGCH to the UE form the non-serving RLS by definition [43].

The cells which are in the serving RLS can issue E-RGCH commands to raise, hold and lower the current UE serving grants. However, the cells in the Non-Serving RLS can only issue "HOLD" or "DOWN" commands. This is a kind of control measure which informs the current cell about an overloading situation at neighbor cells. Therefore, out of all E-RGCH commands, the "DOWN" command has the highest priority and the UE must reduce its serving grants regardless of any other grants it receives. The "HOLD" command has the second highest priority and lastly the "UP" command has the lowest priority to increase the UE's serving grants if no other command is received. Although the 3GPP standards define how the network communicates a serving grant to a UE, the algorithm by which the network determines which commands should be sent on the E-AGCH/E-RGCH is not defined and is left to the mobile network operators [61].

2.2.5.3   *UE Status Report for Scheduling*

The measurement reporting functionality is defined in the 3GPP standards to allow the UEs to communicate their current status. UE status reporting takes two forms, scheduling information transmitted on the E-DCH along with the user data, and a "happy" bit transmitted on the E-DPCCH channel. The scheduling information provides an indication of how much data is waiting to be transmitted in the UE and how much additional network capacity the UE could make use of. For example if the UE is already transmitting at full power then it would be a wastage of resources to increase its serving grant as the UE would be unable to make use of the additional power [60].

The other status reporting mechanism is the "happy" bit information. This is a single bit that is transmitted on the E-DPCCH physical channel. A UE considers itself to be unhappy if it is not transmitting at maximum power and it cannot empty the transmit buffer with the current serving grant within a certain period of time. The period of time is known as the Happy Bit Delay Condition (HBDC) and is signaled by the RRC layer during call setup. Thus the "happy" bit is a crude indication of whether the UE could make good use of additional uplink power.

2.2.5.4   *Uplink HARQ Functionality*

The basic functionality of HARQ is described in section 2.2.2. The HARQ scheme runs in the Node-B for the uplink transmission. The functionality is similar to the HARQ in HSDPA. There are 8 HARQ processes that run in parallel for 2 ms TTI and for each connection. Each time when the UE transmits, the receiving HARQ process in the Node-B will attempt to decode the transport block. If the decoding is successful, the Node-B transmits an ACK to the UE over the E-HICH channel and that HARQ process in the UE will advance to the next transport block. If the decoding of the transport block fails then the Node-B transmits a NACK to the UE on the E-HICH. The UE retransmits the transport block until the maximum number of retransmissions is met. After reaching the maximum number of retransmissions, the HARQ process which runs in the UE will advance to the next transport block. The UE will either use chase combining which means the transmission of exactly the same bits again or incremental redundancy which is a transmission of a different set of bits, depending on how the RRC layer configured the link at call setup.

2.2.5.5   *HSUPA Protocol Architecture*

Figure 2-7 shows the UTRAN protocols which are used for the uplink transmissions. There are uplink related specific layers which are added to the standard UMTS layered architecture [61]. They are the MAC-es/e at the UE entity, the MAC-e at the Node-B and the MAC-es at the RNC. In the UE MAC-es/e are considered as one single layer whereas in the network side MAC-e and MAC-es are considered as separate layers. The new E-DCH transport channel connects up to the new MAC-e, MAC-es and the MAC-es/e layer. The MAC-es/e in the UE contains the HARQ processes and it performs the selection of the uplink data rate based on maintaining the current serving grant and also provides the status reporting. Further this layer creates a transport block based on the scheduling grants received by the MAC-e layer in Node-B. The latter layer contains the HARQ processes, some de-multiplexing functionality and the fast scheduling algorithm. The MAC-es layer in RNC primarily provides reordering, combining and also disassembly of MAC-es PDUs into individual MAC-d PDUs.

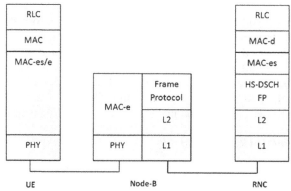

Figure 2-7: UTRAN protocol architecture for the uplink

Since HSUPA supports soft handover, it is possible to receive more than one MAC-e PDU at the RNC from the same UE via different routes (via different Node-Bs). This results in duplicate PDU arrivals. The MAC-es detects such duplicates and delete them before sending the PDUs to the upper layer in the RNC. Further, due to the parallel nature of the HARQ processes it is also possible for MAC-e PDUs to arrive out of order at the MAC-es layer in the RNC. Therefore, the latter layer also does the reordering and provides in-sequence delivery to the upper layer.

2.3 UMTS Transport Network

The UMTS transport network mainly consists of the Iub interface which connects the Node-B with the RNC. Often this network is also named Transport Network Layer (TNL) in scientific reports. Asynchronous Transfer Mode (ATM) is the primary technology introduced in UMTS 3GPP Rel'99 (Rel'99) and release 4 (Rel'4). The ATM technology provides several service priorities which support different QoS requirements of various traffic types and it achieves a very good multiplexing gain for bursty traffic. With the improvement of the QoS support and the transport capacity requirements from Rel'5 (Rel'5) onward, other transport technologies such as IP over DSL have been introduced into the standards. Therefore apart from the main ATM technology, DSL based transport technology is also described in this chapter.

The Iub interface allows the RNC and the Node-B to negotiate about radio resources. It is the most critical interface in the UTRAN from the terrestrial transport network point of view. The designing and dimensioning of this expensive Iub network should be done as cost effectively as possible. Therefore traffic over this network is controlled in order to provide optimum utilization while achieving the required Quality of Service (QoS) guarantees for each service. The trade-off between Optimization of bandwidth (low cost transmission) and provision of QoS is a major challenge for Mobile Network Operates (MNOs).

2.3.1 ATM Based Transport Network

Due to the tremendous growth of Internet and multimedia traffic at the start of 3G mobile communications, scientific research focuses on finding cost effective solutions. ATM (Asynchronous Transfer Mode) is one of the key technologies which are used for high speed transmissions. ATM is designed as a cell switching and multiplexing technology to combine the benefits of circuit switching and packet switching techniques [62]. The term "cell" in the context of ATM means a small packet with constant size. Circuit switching provides constant transmission delay and guaranteed capacity whereas packet switching provides high flexibility and a bandwidth efficient way of transmission. Due to the short fixed length cells transmitted over the network, it can be used for the traffic integration of all services including voice, video and data.

The basic format of the cell which has the size of 53 bytes [62] is shown in Figure 2-8. The cell consists of five bytes header and 48 bytes user or control data. Two different header code structures can be defined in the header depending on the transmission: User Network Interface (UNI) and Network Node Interface (NNI).

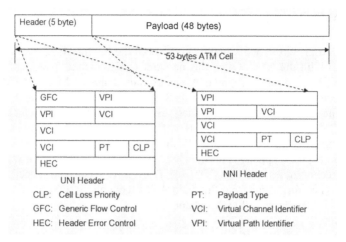

Figure 2-8: ATM PDU format along with the header details

For the purpose of routing cells over the network, VPI (Virtual Path Identifier, 8 bits) and VCI (Virtual Channel Identifier, 16 bits) are defined in the header. The Payload Type (PT) is used to identify the type of data – control or user data – whereas the CLP bit field is used to set the priority. When congestion occurs, the cell discarding technique is applied based on the priority assigned to the cell by the CLP field. For example, packets with CLP = 1 are discarded first while preserving CLP = 0 packets. Finally 8 bits are assigned to the HEC field to monitor header correctness and perform single bit error correction.

After the ATM connection has been set up, cells can be independently labeled and transmitted on demand across the network. Therefore the ATM layer can be divided into VP and VC sub-layers [62] as shown in Figure 2-9. The connections supported at the VP sub-layers, i.e. the Virtual Path Connections (VPC) do not require call control, bandwidth management or processing capabilities. The connection at the VC sub layer, i.e. the Virtual Channel Connection (VCC) may be permanent, semi-permanent or switched. The switched connections require signaling to support establishment, tearing down and capacity management. The permanent and semi-permanent virtual paths are denoted as PVPs and SPVPs throughout this thesis. ATM technology is intended to support a wide variety of services and applications. The control of ATM network traffic is fundamentally related to the ability of the network to provide appropriately differentiated Quality of Service (QoS) for network applications. A set of six service categories is specified. For each one, a set of parameters is given to describe both the traffic presented to the network and the Quality of Service (QoS) which is required from the network. A number of traffic control mechanisms are defined which the network may utilize to meet the QoS objectives.

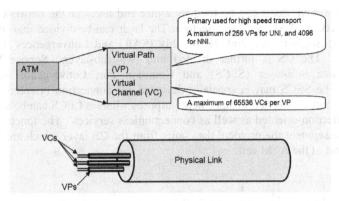

Figure 2-9: VP and VC sub-layers details

The six service categories which are specified in the ATM forum 99 release are as follows.

- Constant Bit Rate (CBR),
- Real time Variable Bit Rate (rt-VBR),
- Non-Real time Variable Bit Rate (nrt-VBR),
- Unspecified Bit Rate (UBR),
- Available Bit Rate (ABR),
- Guaranteed Frame Rate (GFR).

Service categories are distinguished as being real time or non-real time. The categories belonging to real time are CBR and rt-VBR support real time application services depending on the traffic descriptor specifications such as Peak Cell Rate (PCR) or Sustainable Cell Rate (SCR). The other four services categorize the support of non-real time services under the requirements of the traffic descriptor parameters. All service categories except GFR apply to both VPCs and VCCs. GFR is a frame aware service category which can only be applied to VC connections since frame delineation is usually not visible at the virtual path level.

2.3.1.1   ATM Adaptation Layer

When considering the OSI reference protocol architecture ATM belongs to the data link layer. In general, services or applications cannot be mapped directly to ATM cells. This is done through the ATM adaptation layer (AAL). The AAL protocols perform functions of adapting services to the ATM layer and are also responsible for making the network behavior transparent to the application. They represent the link between particular functional requirements of a service and the generic service-independent nature of ATM transport. Depending on the type of service, the AAL layer can be used by either end users or the network. Figure 2-10 depicts the AAL layer architecture. The layer can be divided into two main sub-layers: Segmentation and Reassembly (SAR) and Convergences Sub-layer (CS) [62]. The CS is further divided into two sub-layers: Service Specific Convergence Sub-layer (SSCS) and Common Part Convergence Sub-layer (CPCS). The SSCS part is specially designed for connection oriented services which support connection management purposes whereas CPCS can be shared by both connection-oriented as well as connectionless services. The function of the SAR is to segment the protocol data units from the CS layer which are fitted to the payload of the ATM cell.

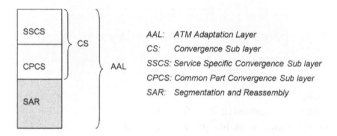

Figure 2-10: AAL architectural overview

There are four types of AAL protocols defined by the ITU standard for the ATM network [ITU I.363]: AAL1, AAL2, AAL3/4 and AAL5 [62]. AAL1 is used to support CBR connections over the ATM network on a per user basis; it is not optimized for bandwidth efficient transmissions. AAL3/4 is an obsolete adaptation standard used to deliver connectionless and connection oriented data over the ATM network. It has a substantial overhead consisting of sequence numbers and multiplexing indicators, and it is rarely used in practice. AAL5 has been developed by the data communication industry; it is optimized for data transport. The AAL2 protocol is the enhanced version of AAL1 which is designed to overcome the limitation experienced from AAL1 and also used in practice. AAL2 is also the adaptation protocol used by the UTRAN Iub interface in the UMTS implementation. This is the most suitable transport layer protocol for real time, variable bit rate services like voice and video. Therefore more details about the architecture and packet formats of the AAL2 protocol are discussed in the next section.

AAL2 is an adaptation layer that is designed to multiplex more than one low bit rate user data stream on a single ATM virtual connection. This AAL provides bandwidth-efficient transmission of low-rate, short, and variable length packets in delay sensitive applications. In situations where multiple low Constant Bit Rate data streams need to be connected to end systems, a lot of precious bandwidth is wasted in setting up conventional VCs for each of the connections. Moreover, most network carriers charge on the basis of number of open Virtual Connections, hence it is efficient both in terms of bandwidth and cost to multiplex as many of these as possible onto a single connection.

2.3.1.2 ATM Qos and Traffic Descriptor Parameters

A set of parameters are negotiated when a connection is set up on ATM networks. These parameters are also used to measure the Quality of Service (QoS) of a connection and quantify end-to-end network performance at the ATM layer. The QoS parameters are described as follows.

- Cell Transfer Delay (CTD): The delay experienced by a cell between the first bit of the cell transmitted by the source and the last bit of the cell received by the destination. Maximum Cell Transfer Delay (Max CTD) and Mean Cell Transfer Delay (Mean CTD) are used.
- Peak-to-peak Cell Delay Variation (CDV): The difference of the maximum and minimum CTD experienced during the connection. Peak-to-peak CDV and Instantaneous CDV are used.
- Cell Loss Ratio (CLR): The percentage of cells that are lost in the network due to error or congestion and are not received by the destination.

The ATM traffic descriptor is defined by four parameters which are

- Peak Cell Rate (PCR): The PCR specifies an upper bound on the rate at which traffic can be submitted on an ATM connection. By knowing this parameter, the network performance objectives can be ensured at the design stage by allocating sufficient transport resources to the network.
- Sustainable Cell Rate (SCR): The SCR is an upper bound on the average rate of the conforming cells of an ATM connection. Defining this parameter is also a good measure to ensure the performance objective of the network. However, this depends on the type of network and the QoS criteria.
- Burst Tolerance (BT): The maximum burst size that can be sent at the peak rate.
- Maximum Burst Size (MBS): The maximum numbers of back-to-back cells that can be sent at the peak cell rate. MBS is related as follows with the other parameters.
- Burst Tolerance = (MBS-1) (1/SCR − 1/PCR).
- Minimum Cell Rate (MCR): The minimum cell rate is the rate at which a source is always allowed to send. MCR is useful for network elements in order to allocate bandwidth among the connections.

2.3.1.3   *ATM Traffic Control and Congestion Control*

The objectives of traffic control and congestion control for ATM are supporting a set of QoS parameters and classes for all ATM services and minimize network and end-system complexity while maximizing network utilization. Designing a congestion control scheme appropriate for ATM networks is mainly guided by scalability and fairness criteria. On the other hand, the design of the network is done in such a way that it should react to congestion in order to minimize its intensity, spread and duration. Three main factors have to be examined for the congestion control: burstiness of the data traffic, the unpredictability of the resource demand and the large propagation delay vs. the large bandwidth. Different levels of network performance may be provided on connections by proper routing, traffic scheduling, priority control and resource allocation to meet the required QoS for the connections.

2.3.2   DSL Based Transport Network

The Digital Subscriber Line (DSL) technology is one of the key technologies used in modern data communication. It uses existing twisted-pair telephone lines (Plan Old Telephony System, POTS) to transmit high bandwidth data. Commonly it is also called xDSL where "x" distinguishes variants of DSL. Mainly there are two types: Symmetric DSL (also called Single pair High bit-rate Digital Subscriber Loop, SHDSL, [65]) or Asymmetric DSL (ADSL) [64]. The latter has different data rates in downstream and upstream whereas SDSL supports equal data rates in both directions. Both ADSL and SHDSL services provide dedicated, point to point, public network access using existing telecommunication infrastructure on the local loop which is commonly known as "last mile", the connection between the network service provider's switching center and the customer's site.

The twisted-pair copper wire supports the frequency spectrum up to 1MHz for the transmission. However the normal voice telephony signal utilizes the spectrum below 4 kHz whereas the rest is not used. Therefore, DSL technology uses this leftover spectrum almost completely for high-speed data transmissions. In general ADSL shares the BW with voice signals whereas the SHDSL uses the complete spectrum without sharing the spectrum with voice telephony [64 and 65].

When the performance of the DSL technology is considered, there is a number of factors that can affect frequencies in the DSL spectrum differently which reduce the effectiveness of the available BW.

- Attenuation: depends on the length and the gauge of the line. The higher the length is, the higher the attenuation becomes, and also narrowing the gauge increases the attenuation as well.
- Reflection and noise: When bridging the lines, reflection and noise can be introduced.
- Cross-talk: bundling results in cross-talk which also depends on the relative position of the line.
- Radio frequency interference: RF interference can occur from external sources, for example any nearby station which transmits RF.

ADSL uses the DMT (Discrete Multi-Tone, [64]) modulation scheme at the transmitter so that it can effectively cope with the above mentioned impairment conditions. DMT splits the available frequency spectrum into several sub-bands. There are 224 subcarriers in the downstream, each occupying a 4 kHz portion of the spectrum. SHDSL uses the Trellis Coded Pulse Amplitude Modulation (TC-PAM, [65]) to provide adaptive performance over the long links.

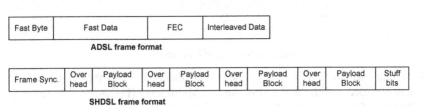

Figure 2-11: ADSL and SHDSL frame formats

The ADSL and SHDSL frame formats are given in Figure 2-11. In addition to the normal frame of ADSL, a super frame is also used for synchronization purposes. After sending 68 normal frames, a super frame or synchronous frame is sent. Each frame starts on a 250 microsecond time boundary; the timing of the frame is kept constant. Depending on prevailing transport conditions, the actual size and content can vary. The Fast Byte field is used to super frame related processing whereas the Fast Data field transmits sensitive data such as audio. The accuracy of the data is ensured by the FEC. Interleaved data is the user data such as Internet data.

The first field of the SHDSL frame is a frame synchronization word used to align the frames. The payload blocks carry the user data; each of them is pre-pended by a corresponding header field for error checking and further synchronization. Each payload block is divided into 12 sub-blocks. The last stuffing bits are also used to support the frame synchronization.

2.3.2.1   *ADSL Network*

Figure 2-12 shows the main entities in the ADSL network [63, 64]. The last mile network, i.e. the network between the switching center and the subscriber's premises is shown along with the relevant components in Figure 2-12.

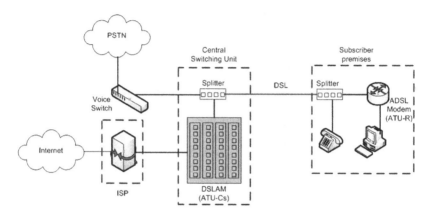

Figure 2-12: ADSL network

The central switching unit which connects to the PSTN and the data network consists of a DSL splitter and a DSL Access Multiplexer (DSLAM). The latter is a collection of ATU-C units where each of them is the termination point of a local loop at the central switching unit. Further it connects to the ATU-R unit (also called ADSL Modem) at the subscriber premises. The splitter is a passive filter which is located at both sides of the local loop. It separates the voice frequencies from the data frequencies when receiving the signals whereas when transmitting it combines the voice and data signals together into one line. The external data network of the Internet Service Provider (ISP) is connected to the central switching unit via the DSLAM; the PSTN network is directly connected via the splitter.

2.4   Long Term Evolution

Traffic demand in modern days is growing exponentially with usage of numerous applications in both real time and best effort domains. In order to be prepared for such demands in the future, 3GPP Rel'8 defines a new mobile broadband access technology often referred to as Long-Term Evolution (LTE) including the development of an evolved core network which is known as System Architecture Evolution (SAE). The bandwidth capability for a UE is expected to be 20 MHz for both the uplink and the downlink [48]. The main targets of these technologies are high peak data rates, low latency, flexible bandwidth support, improved system capacity, reduced UE and system complexities and also reduced network cost. The LTE technology is designed to rely on IP based packet networks. Therefore, the main challenge from the view of mobile network operators is to achieve target data rates and latency by meeting the end user QoS requirements and throughput performance.

2.4.1   LTE Targets

The 3GPP activity regarding 3G evolution fixed the requirements, objectives and targets in the spring of 2005. These targets are documented in 3GPP TR 25.913 [40]. When operating in a 20MHz spectrum allocation, the targets for downlink and uplink peak data-rates are 100Mbit/s and 50Mbit/s respectively [42]. The latency requirements consist of control-plane requirements and user-plane requirements separately. The user-plane latency means the time it spends to transmit an IP packet from the terminal to the RAN edge node or vice versa which is measured on the IP layer. The requirement for the one-way transmission time should not exceed 5 ms [44]. The control-plane latency requires a delay of less than 100 ms.

The global demands for bandwidth increase rapidly and network operators obtain an increasing amount of scattered spectrum, spread over various bands with different bandwidth. Under such circumstances, LTE is targeted to operate in different frequency bands for the spectrum allocation. The 3GPP standardized LTE to operate in channels with 1.25, 1.6, 2.5, 5, 10, 15, and 20MHz bandwidths [47 and 48]. The mobility requirements focus on the speed of the mobile stations. The best performance for LTE is realized at low speeds around 0-15 km/h [43]. However, LTE systems should ensure high performance for speeds above 120km/h as well [43]. The maximum speed which LTE is designed to manage acceptable performance is in the range from 300 to 400 km/h which is mainly dependent on the selected frequency band as well.

2.4.2   LTE Technology

LTE introduces a new air interface technology based on Orthogonal Frequency Division Multiplex (OFDM) which uses closely spaced carriers that are modulated with low data rates. Since carrier spacing is equal to the reciprocal of the symbol period, carriers are orthogonal to each other which leads to a simple equalization effort at the receiver. Therefore, the OFDM based transmission is not only robust against multipath fading effects but also avoids mutual interference [44]. Since due to frequency selective fading only a small number of carriers are affected, the forward error correction is able to effectively recover the corrupted data from the remaining subcarriers. In contrast to the LTE downlink which deploys OFDM, the uplink uses Single-Carrier Frequency Division Multiple Access (SC-FDMA) which provides a lower peak-to-average ratio for the transmitted signal. Saving power in the mobile station in order to keep a long battery life is one of the main considerations to select SC-FDMA in the LTE uplink. The small peak-to-average ratio of the transmitted signal allows a high transmission power, while achieving more efficient usage of the power amplifier in the uplink transmission. The implementation of the modulation and the demodulation functionalities in the LTE uplink and the downlink transmission is computationally efficient due to usage of FFT techniques. The supported modulation schemes for the individual carriers in the downlink and uplink are QPSK, 16QAM and 64QAM [43].

Shared-channel transmission is vital for the LTE transmission technique. In such a scheme, the time-frequency resource which is used as the main shared resource is dynamically allocated among different users. The scheduler is responsible for allocating resources for each user for the downlink transmission. When compared to HSPA channel dependent scheduling, it selects the UE with best the channel condition in the frequency domain whereas the LTE scheduler considers channel variations in both time and frequency domain. It can make decisions to track channel variations as often as every 1 ms with a 180 kHz granularity in the frequency domain [46].

As in HSPA, fast HARQ with soft combining is deployed in LTE as well which enables the UE to request retransmissions to quickly recover erroneously received data blocks. The soft combining of HARQ exploits incremental redundancy to increase the successful decoding capabilities [48].

Another key technique applied in LTE is the multiple antenna technique which is also called Multiple-Input Multiple-Output (MIMO). It can enhance the diversity for reception and transmission. Further, beam-forming can be effectively used at the LTE base station with the usage of multiple antennas.

2.4.3   LTE Architecture and Protocols

The LTE architecture with all network entities is shown as a bock diagram in Figure 2-13. As described above, the LTE network consists of two main parts: Evolved UMTS Terrestrial Radio Access Network (E-UTRAN) and the Evolved Packet Core (EPC). The UE and enhanced Node-B (also referred to as eNB or eNode-B) are the main functional elements which are located in the E-UTRAN whereas the Serving Gateway (SGW), Mobility Management entity (MME) and SGSN are the main functional elements which are located in the EPC. The serving gateway is the main router for an LTE network which connects the E-UTRAN and the PDN (Packet Domain Controller) networks.

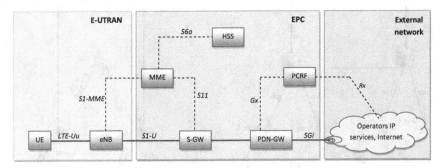

Figure 2-13: LTE architecture with main network entities

From the functional point of view, it routes and forwards packet data to the external network via the PDN. Further it acts as the mobility anchor for the inter-eNB handovers at the user-plane and also for other handovers between the LTE network and the other 3GPP based broadband wireless networks. The MME is the key control element of the LTE access network at the control-plane. Many control-plane functionalities, such as activation and deactivation of bearers, UE tracking and paging procedures, authentication and security key management of UEs, providing initial user identity and selecting the SGW during handovers are performed by this node. Further, Non-Access Stratum (NAS) signals terminate at the MME node as well [43]. The PDN gateway provides connectivity between the LTE network and the external packet data network and performs policy enforcement, packet filtering for each user, charging support, lawful interception and packet screening.

2.4.3.1   E-UTRAN Architecture

The E-UTRAN has a simple flat architecture compared to the HSPA UTRAN architecture and consists only of two main network elements UE and eNB, whereas the UTRAN consists of three main network entities UE, Node-B and RNC [40]. The comparison between the two architectures UTRAN and E-UTRAN is shown in Figure 2-14.

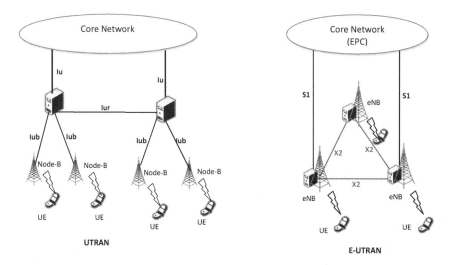

Figure 2-14: UTRAN (left) and E-UTRAN architectures

The eNB in LTE is directly connected with the core network whereas the Node-B in the UTRAN connects through the RNC to the core network. The RNC is the main control entity in UTRAN network, however in the E-UTRAN, this entity is dropped and most of its functionalities are distributed among the enhanced Node-B entity, the MME and the serving GW. The figure shows that the E-UTRAN and UTRAN connectivity to their core networks are done via different interfaces as well. For an E-UTRAN network, the UE is connected to the eNB via the LTE Uu interface and eNB is connected to the EPC via the S1 interface. The additional X2 interface connects the eNBs with each other. All interfaces of the UTRAN have been discussed in the UMTS network architecture.

2.4.3.2   LTE User-Plane Protocol Architecture

The LTE user plane architecture mainly consists of the LTE access network and the core network. However in order to have a better overview, Figure 2-15 shows the end-to-end user-plane protocol architecture from the application layer of the LTE mobile terminal to the application layer of the remote server. Even though it is called user-plane protocol architecture, all application based signaling such as SIP, RTCP and SDP are also routed via this network. The complete architecture uses IP as the transport network protocol. Depending on the services, different end user protocols are involved during data transmissions between end-nodes. For example, best effort traffic uses TCP as the transport protocol whereas the most of the real time applications use UDP.

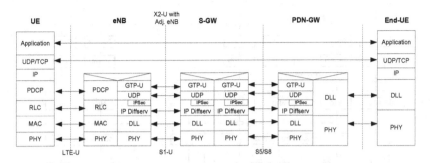

Figure 2-15: LTE user-plane protocol architecture

As shown in Figure 2-15, S1, X2 and S5/S8 interfaces use the 3GPP specific tunneling protocol, called the GPRS Tunneling Protocol (GTP). The protocol stack of these interfaces consists of the GTP-user plane (GTP-U), the transport layer based on the User Datagram Protocol (UDP), the IP based network layer, the data link layer and the PHY layer. Transport bearers are identified by the GTP tunnel endpoints and IP based information such as source address, destination address etc. Apart from routing the user-plane data, IP transport performs service differentiation at the E-UTRAN network entities in order to meet the end user QoS requirements. Further security protection is also provided for user-plane data at this layer using the IPSec security protocol according to the IETF in RFC 2401 [66].

The LTE radio interface between UE and eNode-B is called Uu interface. This interface consists of four main protocol layers in order to transfer the data between eNB and UE securely. They are the Packet Data Convergence Protocol (PDCP) layer, the Radio Link Control (RLC) layer, the Medium Access Control (MAC) layer and the Physical (PHY) layer.

The PDCP layer processes the IP packets in the user plane. Each radio bearer has one PDCP entity that performs header compression/decompression, security including mainly integrity/verification and ciphering/deciphering, and also reordering during handovers. Further details about the PDCP protocol can be found in 3GPP [49].

According to 3GPP [45], the RLC also has one entity per radio bearer and performs a reliable transmission over the air interface, segmentation and reassembly based on the transport block size allowed by the radio interface and also in-sequence packet delivery to the upper layers. The in-sequence delivery is required due to the HARQ process in the MAC layer which can receive out of order arrivals due to retransmissions.

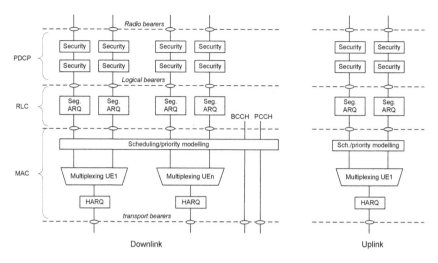

Figure 2-16: Uu protocols functionalities for the UL and the DL

The MAC layer provides one MAC entity per UE, not per radio bearer. The MAC layer provides data multiplexing of different radio bearers which belong to the same UE. It indicates the transport block size to the RLC bearers by considering QoS aspects of the services. There are two very important functional units which are located in this layer: the MAC scheduler and the HARQ processes. The MAC scheduler is in the MAC layer of the eNB. The scheduler allocates the radio resources for each UE based on the agreed QoS criteria. For this process, the user buffer occupancy and service priorities are taken into account. The common set of guidelines which should be agreed by all vendors for scheduling is given in 3GPP [48], however the implementation is vendor specific. The latter may apply algorithms such as priority schemes for different services within the given frame work of the 3GPP specification. As in HSPA, the HARQ processes provide efficient error recovery at the MAC layer which increases the spectral efficiency for the cost of additional delay due to retransmissions.

The physical layer transmits data over the radio channel using the Orthogonal Frequency Division Multiplex (OFDM) and Multiple Input Multiple Output (MIMO) technologies. In the downlink, Orthogonal Frequency Division Multiple Access (OFDMA) allows multiple users to access the medium on a sub carrier basis. In the uplink, Single Carrier Frequency Division Multiple Access (SC-FDMA) is used for the medium access. Extensive details about modeling the PHY layer are provided by 3GPP [47].

2.4.3.3   Control-Plane Protocol Architecture

The control-plane protocol stack between UE and MME and also for the X2 interface is shown in Figure 2-17. This protocol stack includes most of the protocols which are described in the user-plane. From the functional point of view they work in the same manner with one exception at the PDCP layer. In this case, there is no PDCP header compression functionality for the control plane.

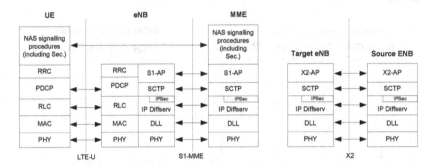

Figure 2-17: LTE C-plane protocol architecture for S1 and X2 interfaces

The Non-Access Stratum (NAS) is the key application layer protocol which provides signaling between UEs and MMEs that are not processed by the eNB. NAS messages are encapsulated into the Radio Resource Control (RRC) protocol and the S1-Application Protocol (S1-AP) in order to provide direct transport of NAS signaling between the MME and the UE. The eNB is responsible for mapping NAS messages between RRC and S1-AP protocols. NAS supports authentication, tracking area updates, paging, Public Land Mobile Network (PLMN) selection and bearer management (EPS bearer establishment, modification and release). The RRC provides the means to transfer NAS signaling information between UE and eNB entities whereas the S1-AP is responsible for transferring signaling information between the eNB and MME over the S1-MME reference point. The S-AP is carried using the Stream Control Transmission Protocol (SCTP) [41]. Furthermore the S1-AP supports QoS bearer management, S1 signaling bearer management, S1 interface management, paging signaling, mobility signaling and tracking area control signaling.

The X2-Application Protocol (X2-AP) is used to transfer signaling information between neighboring eNBs over the X2 interface. It is also carried over the SCTP protocol by providing handover signaling, inter-cell RRM signaling and X2 interface management functionalities.

2.4.3.4   Logical Channels and Transport Channels

The data is exchanged between the MAC layer and the RLC layer using the logical traffic channels. The dedicated traffic channel (DTCH) is used to transmit dedicated user data for both uplink and downlink directions whereas the Multicast Traffic Channel (MTCH) is used to transmit MBMS services for the downlink. The transport channel is used to transfer data between the MAC layer and the PHY layer.

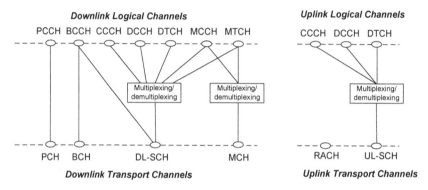

Figure 2-18: Mapping between logical channels and transport channels

Data is multiplexed into transport channels depending on air interface bearer allocation. There are four transport channels for the downlink. They are the Downlink Shared Channel (DL-SCH) which is used to transport user data, the Broadcast Channel (BCH) which is used to transport system information for the DL-SCH channel, the Paging Channel (PCH) which is used for paging information and the Multicast Channel (MCH) which transports some user and control information. The mapping between logical channels and transport channels for the downlink and the uplink is shown in Figure 2-18.

There are two transport channels which are used for the uplink: the Uplink Shared Channel (UL-SCH) is used to transport user data over the uplink and the Random Access Channel (RACH), is used for network access if the UE does not have accurate time synchronization or allocated uplink resources.

2.4.3.5   LTE MAC Scheduler

The LTE MAC scheduler is located in the eNB. Radio resources at the Uu interface are scarce and have to be distributed among the users in the cell in order to achieve optimum throughput while meeting QoS requirements for different services. Due to many factors, such as throughput, fairness and limited resources, designing an appropriate scheduler is a real challenge for mobile network operators. Many advanced PHY technologies such as OFDM provide great flexibility and support such achievements in LTE. The scheduler implementation itself is not standardized but signaling to support scheduling is standardized by the 3GPP specification.

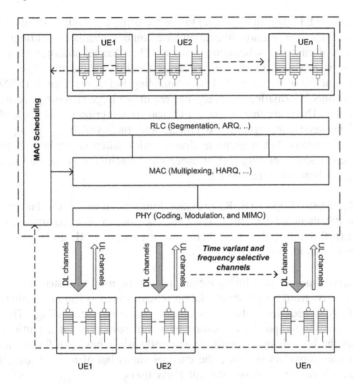

Figure 2-19: LTE MAC scheduler overview

The basic resource unit is the Physical Resource Block (PRB) which is defined in both the time and the frequency domain. In the latter, it has a size of 180 kHz and in the time domain, a sub-frame or transmission time interval (TTI) is 1 ms for both the uplink and the downlink. The radio resources in a cell are dynamically allocated to the UEs in terms of resource blocks through UL-SCH and DL-SCH channels in the uplink and the downlink respectively. Such resource allocation for individual users is performed by mainly considering three factors: current channel condition, UE buffer status (buffer status report, BSR) and QoS requirements. Figure 2-19 shows the scheduler overview in eNB.

Since OFDM is used as the PHY technology, there is great flexibility of selecting frequency blocks which have an optimum channel performance for certain UEs and thus provide higher spectral efficiency for the selected frequency range in which high modulation schemes such as 64 QAM can be deployed. In this manner, the scheduler can select users who provide the best channel conditions in different frequencies and utilize the complete frequency band effectively for scheduling among the selected users.

This means the LTE scheduler is able to use the benefit of multi-user diversity by using frequency-time resource allocation based on best channel conditions among active users during the scheduling process. The channel information of each connection is provided by the UE Channel Quality Indicators (CQIs) for the downlink. For the uplink, the eNB uses Sounding Reference Signals (SRSs). To provide channel information for every TTI requires a large signaling overhead for the scheduler. Therefore this is again a critical design criterion which leads to a trade-off between the signaling overhead and the availability of up-to-date channel information. When accurate channel information is available before the scheduling decision, optimum performance can be achieved at the cost of higher signaling overhead and vice versa.

The buffer Status Report (BSR) provides information about the buffer status, including details about the filling level for each bearer. For the uplink, the UE sends the BSR report to the eNB whereas for the DL, the eNB itself gets the BSR information from its PDCP layer.

The different types of scheduling techniques can be used to distribute resources among the users from fully fair scheduling schemes such as Round-Robin (RR) to opportunistic scheduling schemes such as exhaustive scheduling. The latter optimizes the throughput by allocating even full resources for a particular UE which has the best channel conditions currently while having a full buffer. The fair scheduler considers not only the channel status but also the distribution of available radio resources among current active users.

All above listed facts can be used to design an effective MAC scheduler. The design of the LTE scheduler and the selection appropriate parameters are completely vendor specific upon on their individual goals.

2.4.4   Quality of Service and Bearer Classification

In general, QoS is the concept of providing particular quality guarantees for a specific service. By nature, different applications need different QoS requirements. One user may run more than one application at any time in the UE device or the mobile. The network should guarantee the required QoS for each application individually which is a real challenge for mobile network operators. To cater such demands, LTE groups the services into different service categories and defines a common framework to differentiate such services in the packetized network [45].

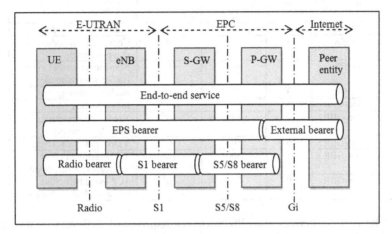

Figure 2-20: EPS bearer mapping in the E-UTRAN.

In general, all applications can be divided into two main categories: real time applications and best effort applications. Real time applications such as VoIP require stringent delay and delay variation (jitter) guarantees whereas the best effort applications such as web browsing and FTP require low packet loss rate guarantees. In a similar manner, the LTE network also distinguishes two types of bearers in a broader view, which are GBR and non-GBR bearers. GBR bearers provide a guaranteed data rate within an acceptable packet delay budget whereas non-GBR bearers are concerned about the reliable packet delivery by providing low packet loss rates. In order to support multiple QoS requirements, Evolved Packet System (EPS) in LTE introduces the EPS bearer concepts which establish the bearer connection between UE and PDN gateway (P-GW). Within the LTE network, the EPS bearer is realized by three subsidiary bearers: S5/S8 bearer, S1 bearer and radio bearer. Figure 2-20 shows the overall QoS mapping and bearer mapping between end-to-end entities through the LTE network [47].

The end-to-end services are defined between end networks, in this case between the LTE UE and the external network. The end-to-end QoS requirements of application flows are realized by the EPS bearers in the LTE network. As shown in Figure 2-20, the EPS bearer consists of three bearers: S5/S8 bearer between P-GW and S-GW, the S1 bearer between the S-GW and the eNB and radio bearer between eNB and UE in the LTE network. All these bearers are mapped one-to-one through each interface and keep a unique bearer identity to guarantee the QoS targets through the networks.

Table 2-1: LTE QoS classes defined by 3GPP

QCI	Resource type	Priority	Packet delay budget (ms)	Packet error loss rate	Example services
1	GBR	2	100	10^{-2}	Conversational video
2	GBR	4	150	10^{-3}	Conversational video (Live streaming)
3	GBR	5	300	10^{-6}	Non-conversational video (buffered streaming)
4	GBR	3	50	10^{-3}	Real time gaming
5	Non-GBR	1	100	10^{-6}	IMS signaling
6	Non-GBR	7	100	10^{-3}	Voice, live streaming, interactive gaming
7	Non-GBR	6	300	10^{-6}	Video (buffer streaming)
8	Non-GBR	8	300	10^{-6}	TCP based best effort applications (web, email)
9	Non-GBR	9	300	10^{-6}	TCP based best effort (a large file downloads)

From the view of services, the EPS bearer is a sequence of IP flows with the same QoS profile established between the UE and the P-GW. There are five QoS parameters which are defined by 3GPP standards [46]. They are

- QoS class identifier (QCI),
- Allocation and retention priority (ARP),
- Guaranteed bit rate (GBR),
- Maximum bit rate (MBR),
- Aggregate maximum bit rate (AMBR).

Each of these QoS parameters provides different tasks in the IP based packet network. The QCI is a scalar value which is used for bearer level packet forwarding over the network whereas ARP supports admission control procedures for the connection establishment and release functionalities based on congestion in the radio network. As the definition states, the GBR parameter provides guaranteed bit rates for the service and is only used for GBR bearers. The MBR parameter specifies the maximum bit rate which can be provided by the bearer whereas AMBR specifies the limit of the aggregate bit rate that can be expected to be provided by all non-GBR bearers associated with a PDN connection (e.g. excess traffic may get discarded by a rate-shaping function). An end application is specified with QoS parameters before packets are sent to the IP based packet network. The QCI is one of the important parameters which provide information about how the packet of a particular service type should be treated in the packet network concerning QoS. The QCI value is mapped into Differentiated Service Code Points (DSCP) and is included in the Type of Service (ToS) field of the IP packet header. Therefore, any IP packet which traverses the network carries the

QoS information along with the packet itself. When a network node processes the packets, it can differentiate packets regarding QoS and can provide the required QoS guarantee for a particular service flow. In order to handle traffic uniformly throughout the LTE network, some of the QCIs are standardized by 3GPP. Therefore all vendors and mobile network operators have to agree on these service requirements. They are listed in Table 2-1. For each QoS flow, a separate EPS bearer is required to be established between the UE and the PDN gateway. Therefore IP packets must be categorized into different EPS bearers. This is done based on Traffic Flow Templates (TFTs) which use the IP header information such as ToS, TCP port numbers, source and destination addresses to select packets from different service flows such VoIP and web browsing.

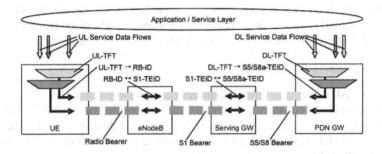

Figure 2-21: TFT based packet filtering procedure defined by 3GPP [46]

This TFT based packet filtering procedure is defined by 3GPP [46] and is shown in Figure 2-21 for the uplink and downlink service data flows separately. When the UE is attached to the LTE network, it is assigned an IP address by the P-GW by establishing at least one EPS bearer. This bearer is called "default" EPS bearer and it is kept throughout the life time of the connection with the PDN network. Since the default EPS bearer is permanently established, it always has to be a non-GBR bearer. Any additional EPS bearer which is established by the UE is called "dedicated" EPS bearer. Depending on different application flows, any number of dedicated bearers can be established between the UE and the P-GW at any time.

2.4.5   LTE Handovers

LTE is designed to provide seamless mobility. UE handovers occur with imperceptible delay and an acceptable number of packet losses which can be recovered by the upper layers. In contrast to HSPA there is no centralized controller in LTE [46]. All centralized functions are performed at the eNB. Therefore soft handover is not possible for LTE. Whenever UE performs a handover, buffered data in eNB has to be forwarded to the corresponding cell in eNB. Data protection during the HOs is handled by the PDCP layer. UE mobility can be categorized into intra-LTE mobility and inter-RAT (Radio Access Technologies, such as UMTS and CDMA2000) mobility. During this work only intra-LTE mobility is considered, which occurs either between eNBs or between cells. Based on this, handovers within E-UTRAN can be also categorized into two types: inter-eNB handovers which occur between two eNBs and intra-eNB handovers which occur between cells within the same eNB. In both cases the main data handling is performed by the eNB. For the first case, the buffered data in the source eNB where the UE is currently located has to be forwarded via the X2 interface to the target eNB where the UE is moving to. In this case data should be protected and a fast transmission between two eNBs should be executed. Not only the data but also the load or interference related information also has to be forwarded to the target eNB. In the latter case, the eNB handles the data and signaling information within its own entity and forwards data to the target cell from the source cell based on the new UE connectivity.

2.4.5.1   *Inter-eNB Handovers*

As mentioned above, inter-eNB handovers occur between two eNBs which are called source eNB and target eNB. Since data is forwarded over the X2 interface during HOs, the X2 protocol architecture has to be considered in detail. The X2 user-plane protocol stack is given in Figure 2-22 according to the 3GPP specification [47].

The PDCP layer takes care of the data protection and the GTP establishes a tunnel through the transport network between source and target eNBs. The transport network guarantees the QoS aspects and provides the required priority to the forwarded data. To minimize the packet losses, data forwarding is performed on a per-bearer basis. Minimizing delay for forwarding data is required for the real time applications such as VoIP whereas a very low loss probability is required for best effort applications such as email and web browsing which mainly use the TCP protocol. In the context of an HO, the term "lossless", means a very low loss probability; the HO is performed by the PDCP layer which can use an ARQ scheme. Further, the PDCP layer also provides in-sequence delivery functionalities for forwarded data. Once the UE has established the connection with the target eNB, the MME is notified for path switching procedures by the target eNB. In this case, the Late Path Switching (LPS) or the backward handover is used to minimize the losses and also to minimize the interruption during HOs.

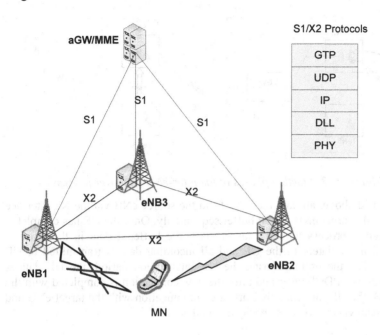

S1/X2 Protocols

GTP
UDP
IP
DLL
PHY

Figure 2-22: X2 protocols and inter-eNB HOs

As shown in Figure 2-23, first the UE is connected to the source eNB (SeNB) and they communicate with each other. Once the UE reaches the tracking area of the target eNB, it initiates an HO by informing the MME via the source eNB. Before the UE disconnects from the source eNB, internal bearers are set up between the target eNB and the source eNB. This is done using inter-eNB signaling over the X2 interface. Then the UE starts a new connection process with the target eNB and meanwhile all data including unacknowledged data is buffered and newly arriving data is forwarded to the target eNB.

The target eNB buffers the forwarded data until the UE finalizes the new connection process with the target eNB. Figure 2-24 shows that the buffered data of the source eNB is forwarded and then incoming data to the source eNB is also forwarded via the X2 interface.

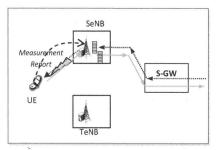 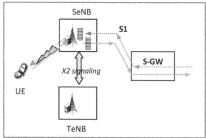

Figure 2-23: Starting phases of the inter-eNB handover process

As mentioned above, all data received from the source eNB via the X2 interface is stored in the target eNB PDCP buffer sequentially. Once the UE has completed the connection process with the target eNB, the buffered data in the latter is transmitted immediately to the UE and all incoming data is transmitted to UE afterwards. For the uplink, during the UE interruption period, the UL data is stored in the UE PDCP layer and after the new connection is completed with the target eNB, the UE immediately starts a communication with the target eNB and sends the data via the S1 to the respective end nodes.

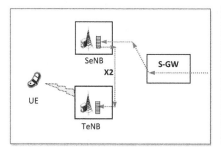 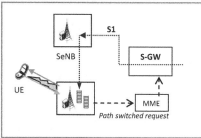

Figure 2-24: Data forwarding during HO over process via X2 interface

For DL path switching, information about the new UE connection to the target eNB is sent to the S-GW via the MME. This is the reason why this method is called late path switching. Once the path is switched to the target eNB, new data arriving from the PDN uses the new path via the S1 interface.

To indicate path switching functionality to intermediate nodes, an end-marker PDU is used. The last PDU before the path switch is the end marker PDU which indicates the end of the transmission via the source eNB. When the eNB receives the end-marker it assumes that the handover is completed and no further data will come along this route. Then the source eNB releases the resources for that UE and passes the end mark transparently to the target eNB. The complete HO procedure according to the 3GPP specification is given in the flow chart in Figure 2-26. The dotted arrows show the control plane signaling whereas the line arrows show the user plane messages and data handling. There are 18 steps which are shown in the flow chart with different signaling messages.

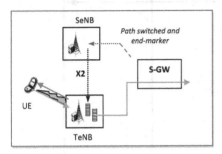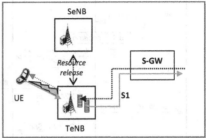

Figure 2-25: Path switched procedures and last stage of HO activities

At step 1, the source eNB configures the UE measurement procedures according to the area tracking information whereas in step 2, the UE is triggered to send a *Measurement Report*. The source eNB makes the decision to handover UE to the target eNB based on the received *Measurement Report* at step 3. Next, the source eNB issues a *Handover Request* message to the target eNB which passes necessary information to prepare the handover at the target eNB. At step 5, Admission Control will be performed by the target eNB dependent on the received radio bearer QoS information and the S1 connectivity to increase the likelihood of a successful handover.

At step 6, the target eNB prepares the handover with L1/L2 and sends a *Handover Request Acknowledge* message to the source eNB. This message includes a transparent container which includes the new C-RNTI and the value of the dedicated preamble to be sent to the UE as part of the *Handover Command*. The source eNB sends an RRC *Handover Command* message towards the UE at step 7.

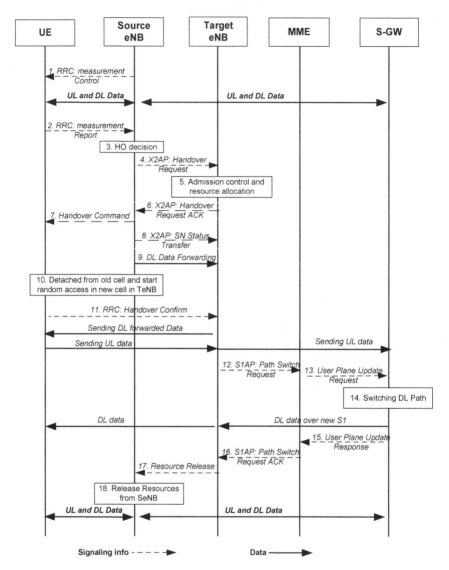

Figure 2-26: Inter-eNB handover procedures

Next the *SN Status Transfer* message is used to transfer PDCP layer information from the source eNB to the target eNB to ensure UL and DL PDCP SN continuity for every bearer that requires PDCP status preservation. Afterwards, DL data forwarding is started from the source eNB to the target eNB.

The *Handover Command* is sent by the source eNB to perform the handover immediately to the target eNB. When the UE has successfully accessed the target cell, it sends the *Handover Confirm* message to the target eNB to indicate that the handover procedure is completed for the UE and starts sending downlink data forwarded from the source eNB to the UE, so the latter can begin sending uplink data to the target eNB as well.

At step 12, the target eNB sends a *Path Switch Request* message to the MME to inform it that the handover has been completed. The new TNL information is then passed to the MME and to the target eNB in step 16. The resources at the S-GW are then released some time after step 16.The MME sends a *User Plane Update Request* message to the S-GW. The S-GW switches the downlink data path to the target eNB and sends a *User Plane Update Response* message to the MME to confirm that it has switched the downlink data path. The MME confirms the *Path Switch Request* message with the *Path Switch Request Ack* message. By sending a *Release Resource* message, the target eNB informs the source eNB of the success of the handover and triggers the release of resources.

2.4.5.2   Intra-eNB Handovers

Intra-eNB handovers are performed among the cells within the same eNB. In this case, there is no signaling with EPC and the HO procedures are very simple compared to inter-eNB HOs as shown in Figure 2-27. All HO procedures are completely done within the eNB. The HO procedure from one cell to another in the same eNB is shown Figure 2-27. Based on the measurement report received from the UE, the eNB takes the decision for the handover to the target cell. Admission Control will be performed in the target cell and the resources are granted by setting up the required radio bearers for that UE. During this HO process DL data is buffered at the eNB and UL data is buffered at the UE. Once the HO process is successful, the UE sends a *Handover Confirm* message to the eNB to indicate that the handover procedure is completed. Then both the UE and the eNB immediately start transmitting data for UL and DL respectively.

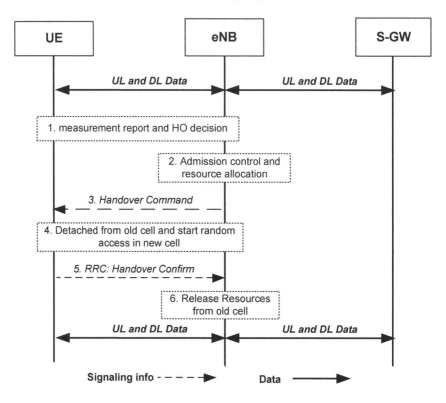

Figure 2-27: Intra-eNB handover procedures

3 HSPA Network Simulator

The implementation of a real-world system is a time consuming and costly working process. Still, after implementing such a complex system, it might not provide the expected behavior or performance at last. This can result in wastage of resources and huge financial downturns for any institution. For this reason, these systems have to be realized with a low probability of risk. Therefore, before the real implementation, there should be ways to estimate the risks of behavior and performance of such systems. To fill this gap, imitation of complex systems can be modeled in simulations and also analyzed with respect to the focused requirements. Understanding the requirements and the objectives are the key elements of designing a simulation model. A model cannot be more accurate than the understanding of the corresponding real-world system. Many approaches such as simulation and analytical considerations are introduced to cater for such real-world scenarios. The analytical approaches are more theoretical and use mathematical modeling. They are efficient in analyzing simplified models. However, today real world systems are very complex having many dependent and independent parameters and cannot be easily described using analytical models. Therefore, in order to analyze the performance and evaluate behavioral aspects, computer based simulation models are introduced. The latter can realize many complex systems and can also evaluate the performance in a wider scope. However depending on the complexity, these systems require longer processing times and longer implementation time than some analytical approaches require.

3.1 Simulation Environment

For the HSPA network design and development, the OPNET software is used as the simulation environment. It is a discrete event simulation tool and provides a comprehensive development environment mainly supporting the modeling of communication networks and distributed systems. It supports a user friendly environment for all phases of system modeling including model design, simulation, data collection, and analysis.

Key features of the OPNET simulator which is specialized in communication networks and information systems are object orientation, hierarchical models, graphical models, flexibility to develop detailed custom models, application program interface (API), application specific statistics, integrated post-simulation analysis tools, interactive analysis, animation and co-simulation. For the design and development of the HSPA network simulator, the basic OPNET modeler is used. It includes the basic communication protocols and their basic functionalities. Further, there are many kernel procedures and dynamic link libraries (DLL) which can be effectively utilized to develop new protocols. Further details about the OPNET simulator are given in the OPNET web page [67].

3.2 HSPA Network Simulator Design

To investigate the performance of HSPA broadband technologies, simulation models have been developed by the author. It is difficult to focus all the aspects of the above technologies in detail for such complex systems. Therefore, the focus of these simulators is concentrated on transport network performance and its Optimization aspects. The end user performance is evaluated by considering the possible effects on the transport network such as congestion, traffic differentiation and handover. As mentioned above, the OPNET tool is used to design the HSPA network simulator. The basic models from OPNET are taken as the basis for the network simulator development. All required new protocol functionalities for transport network investigations have been added on top of these basic models. Further link level simulator traces were used to get the relevant properties of physical layer characteristics for the development of the HSPA network simulator which has the main focus on the transport network and system level performance analysis.

The HSPA network simulator has been designed in two steps. First, the HSDPA part was designed and later the HSUPA part was added. Based on the investigation requirements, the HSDPA or HSUPA units can be used separately or combined. Therefore, the HSPA network simulator has a great flexibility for different aims of investigation. This chapter mainly describes the design and development of the two simulator units as a summary. In addition to the implementation, some of the main features of few protocols are elaborated in more detail.

3.2.1 HSDPA Network Simulator Model Design and Development

A comprehensive HSDPA simulation model has been developed by using the OPNET simulation environment. The main protocols and network structure for this simulator are shown in Figure 3-1 which shows the simplified protocol architecture specifically designed for the transport network based performance analysis. However, it also provides broader coverage of the analysis for HSDPA due to the implementation of all UTRAN and end user protocols. Therefore, in addition to the transport network layer (TNL) performance investigation, it allows Radio Link Controller (RLC), Transmission Control Protocol (TCP) and application layer protocols performance analyzes and thus provides a better overview about the end user performance.

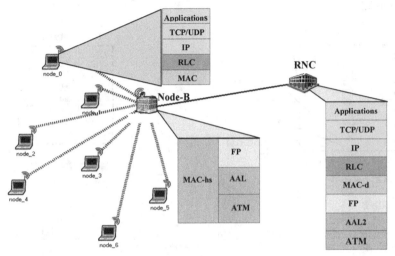

Figure 3-1: Simplified HSDPA network simulator overview

When the proposed network and the real system are compared, many network entities and relevant protocol functionalities are modeled in a simplified manner since they are not the main focus of the investigation. For example, the Internet nodes and the core network protocol functionalities are simplified and added to the RNC node itself. The Radio Access Bearer (RAB) connections are provided by the three upper layers (Application, TCP/UDP and IP) of the proposed RNC protocol stack. They represent the load generated by the core network and the Internet nodes to the transport network. The rest of the protocols in the RNC protocol stack are the transport network protocols, namely Radio Network Control (RLC), Medium Access Controller in downlink (MAC-d), Frame Protocol (FP) and the ATM based transport layers.

The user and cell classification are done in the Node-B. The HSDPA simulator was designed in such a way that one Node-B can support up to 3 cells – each with a maximum of up to 20 users. In addition to the transport layers, the Node-B consists of MAC-hs and FP layers as shown in Figure 3-1. The physical layer of Node-B is not modeled in detail within the simulator. The physical layer data rate for each UE is considered in the MAC layer as a statistical approach and are statistically analyzed and emulated at MAC level as the relevant per-user data rates. With the help of the MAC-hs scheduler, individual user data rates which emulate the wireless channel behavior are scheduled on TTI basis. The number of MAC PDUs is selected for each user from the corresponding MAC-d probability distributions taken from dedicated radio simulation traces. Further details about the HSDPA scheduler implementation and related functionalities are given in section 3.3.

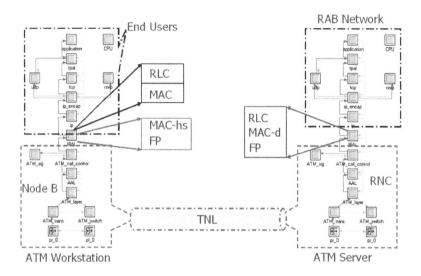

Figure 3-2: Detailed HSDPA network node models' protocol stack

Figure 3-2 shows a more detailed network node model of the HSDPA simulator. As depicted in the figure, the network entities and UTRAN protocols are implemented on top of the OPNET ATM node modules. ATM advanced server and workstation modules are selected from the existing OPNET modeler modules and modified in such a way that they complete the simplified HSDPA protocol architecture. The protocols RLC, MAC-d and FP are added to the ATM server node model. The key Node-B protocols: FP and MAC-hs and the user equipment protocols: MAC and RLC, are implemented in the ATM workstation node model. In order to analyze the effects of the transport network for the end user performance, all above newly added protocols were implemented according to the specification given by 3GPP and ATM forum [22, 62].

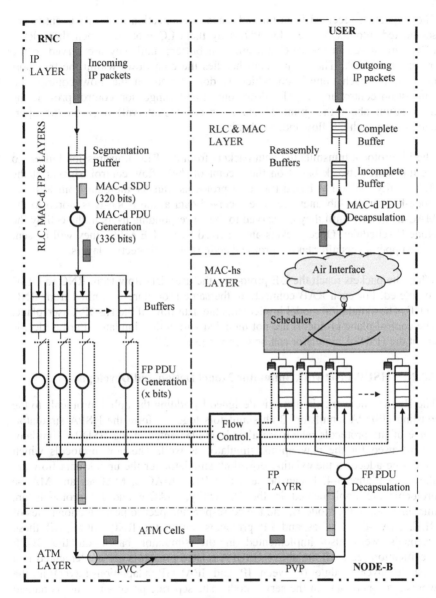

Figure 3-3: HSDPA user-plane data flow

The overview of the detailed implementation of the MAC-d user data flow is shown in Figure 3-3. The main functionalities of RNC and Node-B protocols are shown in this figure. In summary, the user downlink data flow works as follows.

The IP packets received from the radio bearer connection in the RNC entity are segmented into fixed size RLC PDUs by the RLC protocol. Then these RLC PDUs are stored in the RLC transmission buffers until they are served to the transport network. The FP protocol handles the data flows within the transport network for each connection which is done based on the flow control and congestion control inputs. The flow control and congestion control provide the information about the granted data rates (credits/interval) over the transport network for each UE flow independently.

The FP protocol transmits the data packets from the RLC transmission buffers to the transport network based on the receipt of these flow control triggers at the RNC entity. All ATM based transport protocols transfer the data from RNC to Node-B over the Iub interface. The received data at the Node-B is stored in the MAC-hs buffers until they are served to the corresponding users in the cell by the Node-B scheduler. Once packets are received by the UE entity, they will be sent to the upper layers after taking protocol actions at the respective layers.

When the packets reach the UE protocols, the user data flow from RNC to UE is completed. For each RAB connection, the same procedure is applied from RNC to UE. The simulation model implements the data flow of the HSDPA user-plane. The control-plane protocols are not modeled within the simulator. Further details about the HSDPA simulator can be found in [3 and 5].

3.2.2   HSUPA Network Simulator Model Design and Development

The HSUPA network simulator is designed by adding the uplink protocols to the existing HSDPA simulator. Both simulators together form the HSPA simulator. Some of the protocols which are already implemented in the HSDPA simulator can be used for the new uplink simulator as well. The new protocols which should be added to the existing downlink simulator for the uplink data flow are shown in Figure 3-4. It can be seen that RLC, MAC-d, MAC-es and MAC-e protocols are implemented in the UE entity, MAC-e and FP protocols are implemented in the Node-B, and the corresponding peer to peer functionalities of RLC, MAC-d, MAC-es and FP protocols are in the RNC entity. All these protocols were also implemented in the simulator based on the 3GPP specification [22]. All the above HSUPA related protocols are implemented in a separate layer module (between IP and IPAL=IP adaptation Layer) in the workstation node and in the server node. The separate process module is named HSUPA process module or HSUPA layer for the simulator.

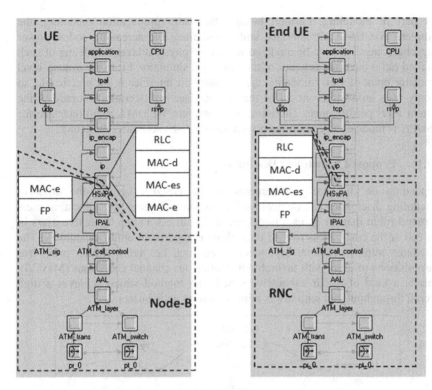

Figure 3-4: Simplified HSUPA protocol stack

The E-DCH scheduler is the main entity which models the air interface functionality in HSUPA. It is implemented in the MAC-e protocol of the Node-B. Further details about the MAC modeling are given in chapter 3.3. So far, a brief overview about the model design and development of HSUPA and HSDPA has been discussed. Providing complete details about the implementation of all protocols in the simulator is not the focus of this dissertation. The idea is to provide a good understanding about the simulators for transport network analysis. However, some details about the key protocols which are directly related to the transport network are elaborated in the next section of this chapter.

3.3 HSDPA MAC-Hs Scheduler Design

The exact modeling of the air interface is a very complex topic. There are several considerations such as propagation environments with short and long term fading as well as interference which is caused by other users of the same cell and of other cells (inter-cell interference). All these effects have to be considered during this part of modeling. Further, besides the radio propagation modeling, the complete W-CDMA air interface with scrambling and spreading codes, power control etc. needs also to be modeled within the physical layer. Modeling of such detailed radio interface is a part of the link level simulator. From the system level simulator point of view, modeling the complete air interface is not practicable, so it only uses an abstract model of the air interface with certain accuracy of the system level performance. Therefore, the complete WCDMA radio interface for HSDPA is modeled in combination with the MAC-hs scheduler modeling.

3.3.1   Proposed Scheduling Disciplines

Depending on the QoS requirements of the HSDPA system, the different scheduling disciplines can be deployed as shown in Figure 3-5. Allocating users in round robin manner gains a high degree of fairness among the active users in the cell at the cost of less overall throughput compared to unfair schedulers. The scheduler employs channel dependent scheduling, i.e. the scheduler prioritizes transmissions to users with favorable instantaneous channel conditions (MaxC/I). This is a kind of unfair exhaustive scheduling method which achieves a high overall throughput, but with lower fairness among active users.

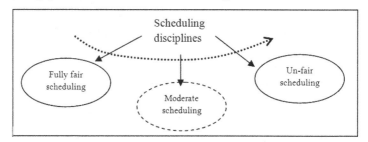

Figure 3-5: Scheduling discipline overview

For the HSDPA TNL analysis, both the round-robin based fair scheduling scheme and the channel dependent based exhaustive scheduling scheme are modeled in the simulator. The per-user throughput over the radio interface is analyzed using the traces which were taken from the radio simulator. The radio simulation output is used as the input to the MAC-hs scheduler. The traces are provided by the radio simulator for 27 cells, each with 20 users. These traces were analyzed and investigated thoroughly to understand the user channel behavior over the radio interface. Since radio traces were taken for 20 users/cell scenarios, the TNL simulation is also performed with a group of 20 users per cell. Scheduling is performed on a per TTI basis and during the investigation of the traces the following key observations have been identified.

- Always two users were scheduled for every TTI and
- those two users occupy a certain number of TTIs consecutively which means the scheduling has a bursty characteristic.

The per-user throughput over the radio interface is represented by the number of MAC-ds per TTI in the traces. Therefore, two probability distributions were derived from the user traces: one gives the probability for the number of MAC-d PDUs per TTI and the other one the number of consecutive TTIs inside a burst. These two probability distributions were used for the scheduling decisions in every TTI.

3.3.2   Round Robin Scheduling Procedure

For the TNL Round Robin (RR) scheduling, at each TTI, two users are selected. At the first TTI, they are chosen independently with two random values for the number of MAC-ds/TTI and burst of consecutive TTIs. These parameters are taken from the probability distributions which are derived from the corresponding UE traces. The scheduled user transmits the data with the amount of the selected number of MAC-ds during consecutive TTIs. The next user who is to be scheduled according to the RR service discipline is selected if either one of the current scheduling UEs elapses it's consecutive TTIs or finishes its data in the transmission and retransmission buffers. Each time when a new user is selected for transmitting, the corresponding number of MAC-ds and burst of TTI are also selected randomly from the probability distribution. This procedure continues until the end of the simulation. The RR scheduler always selects two users for every transmission if they have data in their transmission buffers.

3.3.3   Channel Dependent Scheduling Procedure

The TNL simulator also provides the exhaustive scheduling approach which is based on the best channel quality. In this approach, two best channel quality UEs are selected for the transmission in every TTI. Since there is a direct relationship between the best channel quality and the number of MAC-ds/TTI, two UEs who have the highest number of MAC-ds/TTI are selected for scheduling for each TTI. Selecting the two best UEs is done in following manner. First active users who have data in their transmission buffers are chosen as an eligible user list for the current transmission. Next, the number of MAC-ds and burst TTIs are randomly selected from individual distributions for each of these UEs who are in the eligible list. Within the list, these selected UEs are prioritized based on number of MAC-ds/TTI from higher to lower. Therefore the user who is on the top of the list has the highest number of MAC-d/TTI and also corresponds to the best channel quality for the current TTI. In order to start scheduling, at the very beginning, the two best priority UEs (top two) are chosen from the priority list for scheduling. Whenever one out of two finishes its transmissions or the duration of consecutive TTIs elapses, the next user is selected for scheduling by applying the same procedure again from the beginning. First, the prioritization procedure is started and then the best prioritized user who has the highest number of MAC-ds/TTI is selected.

3.3.4   HSDPA Scheduler Architecture

Figure 3-6 shows the block diagram of the MAC-hs scheduler which contains the MAC-hs transmission buffers and the retransmission buffers. The per-user transmission buffer is the main buffering point in the Node-B. Buffer capacity at the Node-B should be maintained at an adequate level in order to avoid underutilization of radio channel capacities and also to minimize data losses during UE Handovers (HOs).

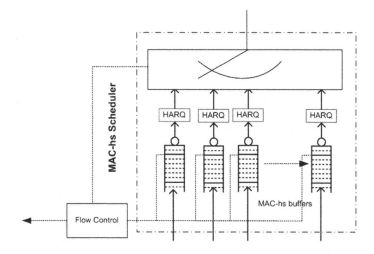

Figure 3-6: Overview of the MAC scheduler and HARQ processes

Flow control is developed for this purpose and the flow over the transport network is controlled based on the changing radio channel capacity. As shown in Figure 3-6, two buffer thresholds are maintained at the Node-B user buffers. More details about the flow control are discussed in chapter 4. In addition to the transmission buffer, each user has a separate HARQ process. More details about the HARQ are discussed later in this chapter.

3.4   E-DCH Scheduler Implemetation

HSUPA uses the E-DCH scheduler for the uplink transmission. This is a scheduler of its own and there is no relation to downlink scheduler functionalities.

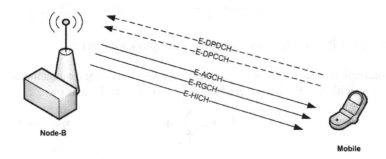

Node-B

Mobile

Figure 3-7: DL and UL information channels for the E-DCH scheduler

It uses, however, downlink control channels such as E-AGCH and E-RGCH to receive the information about the user scheduling grants in the uplink. Further, the uplink control channels indicate what the UE transmits as a "request for transmission", which includes the scheduling information (SI), as well as the happy bit. The E-DCH scheduler is implemented in the Node-B. In the decision making process, it uses QoS information acquired from the RNC and scheduling information gathered in Node-B. A simple overview of the downlink and the uplink control information exchange between UE and eNB is shown in Figure 3-7. The E-DCH scheduler implementation procedures are not specified in the 3GPP specification except some main guide line for the implementation. Therefore, implementation is mostly vendor specific. The detailed HSUPA scheduler implementation is discussed next.

3.4.1   E-DCH Scheduler Design and Development

The HSUPA scheduler decides when and how many UEs are allowed and in which order to transmit during the next transmission opportunity based on their requests and the individual channel knowledge. The UE request mainly includes the amount of data to be transmitted in terms of transmission buffer level information and UE power information. By considering the aforementioned requests and gathered information, the uplink scheduler effectively distributes the available uplink resources among the UEs in every TTI.

The main uplink resource which is taken into consideration is the interference level, or, more precisely, the noise rise (NR) which is mostly in the range of 4dB to 8dB. In a real system, the value of the NR is given by the RNC and changes over time. However for this implementation, the NR is assumed to be a constant value over the whole simulation run. The scheduler uses the "uplink load factor" to maintain the interference level within the cell. A direct relationship between the NR and the uplink load factor is given by equation *3-1* [18].

$$\eta_{UL} = \frac{NR - 1}{NR} \qquad\qquad equation\ 3\text{-}1$$

where the NR is Noise Rise and η_{UL} is the total uplink load factor.

The total uplink load factor can be formulated using individual load factors which are user specific and the interference ratio, α from other cells to its own.

$$\eta_{UL} = (1 + \alpha) \sum_{j=1}^{N} \eta_j \qquad\qquad equation\ 3\text{-}2$$

where η_j represents the load factor for user *j*. This user specific load factor can be derived as follows ([17] [18]).

$$CIR_j = \frac{E_b}{N_0} \times \frac{R_j}{W}$$

$$\eta_j = \frac{1}{1 + 1/CIR_j} \qquad\qquad equation\ 3\text{-}3$$

where CIR is the carrier to interference ratio. It can be taken from prior link level simulations, which depend on the UE speed as well as the BLER. E_b/N_0 is the ratio between the bit energy and noise energy, R_j is the data rate of the jth UE and W is the chip rate of HSUPA which is equal to 3.84 Mcps [18].

The users in the uplink transmission are categorized into two main groups based on their resource grants which they receive from the Node-B: primary absolute grant (PAG) and secondary absolute grant (SAG) UEs. The current implementation always gives the priority to the SAG UEs over PAG UEs. Therefore, first resources are given to the SAG UEs and then the residual resources are allocated to the PAG UEs. The key features which were taken into consideration for the E-DCH scheduler implementation are listed below.

1. The absolute grant (AG) channel carries PAG or SAG of the UE.
2. The relative grants (RG) channel carries "up", "down" and "hold" scheduling information.
3. The scheduler considers RLC buffer occupancies of the UEs for scheduling.

4. The scheduler uses the multi-level SAG grants and the TNL congestion control inputs in order to make scheduling decisions.

The absolute grants for the scheduler are limited due to a limited number of AG channels. Since these channels are downlink channels, they use the same resources that HSDPA uses. During the E-SCH development those aspects are taken into consideration.

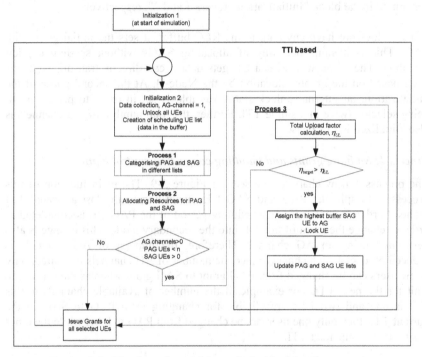

Figure 3-8: The flow chart of the uplink scheduler functionality

So far, the main features and main entities required by the uplink scheduler such as uplink throughput for WCDMA have been discussed. Next the procedure of the uplink scheduler is discussed.

Figure 3-8 shows the main flow chart which describes the overall HSUPA scheduler functionality in detail for every TTI. Each procedure of this flow chart is described separately.

Initialization procedures

At the beginning of the simulation, the target noise rise (NR) of the Node-B is initialized and all the HSUPA users are initialized with minimum SAG (for example 32kbit/s). At the start of each TTI, the data collection phase is also started in which the scheduler collects and updates all required information, mainly the current UE RLC buffer occupancy, the last UE transmission data rate and the congestion control inputs. All these initialization procedures are performed in the block "Initialization process 1 and 2", respectively.

If a UE does not have any data in the RLC buffer, it sets the initial grant with SAG. This is a simplified way of allocating SAGs without sending explicit signaling. Therefore, whenever a UE gets data it can directly transmit with this SAG based on the grants allocation by the Node-B. At the second phase of the initialization, an internal unlock status variable is set in order to protect a UE being selected twice in a single TTI. Further, the total number of AG channels is also initialized in this phase.

Process 1: Splitting grants and handling congestion control input

The process 1 flow chart can be seen in Figure 3-9. The main function of this process is to split the PAG and SAG UEs according to the last allocated TTI grants. Further it checks if the scheduling period of the PAG UE is completed in order to change the grant of this UE into the secondary mode. This change is also performed using an AG channel. Therefore, the number of AG channels is reduced by one at each change. Since the number of AG channels is limited, only a few users can be changed from PAG grant to SAG grant whereas the rest has to wait for the next TTI. For example, if the number of available channels in this TTI is one and two UEs are eligible for changing from PAG to SAG in the current TTI, then only one user can be changed from PAG to SAG while the other has to wait for the next TTI.

Another function inside process 1 is the congestion control action that the scheduler is supposed to take based on the congestion indication. The congestion control is done only against PAG users. If there is a congestion indication due to a PAG user then the scheduler reduces the UE grant by one step down using RG = Down. Even though the congestion control module signals to the scheduler a grant limit for that UE which is more than one step lower than the previous grant, the scheduler cannot reduce the scheduling grants within one step or one TTI. It will decrease the grant in a stepwise manner, one step in each TTI (by using RG = Down) until the grants are equal to the value recommended by the CC module. This CC grant limit is also called Reference Congestion Limit (RCL). At the end of process 1, the scheduler will have two user lists, one corresponds to the PAG UEs and the other corresponds to the SAG UEs.

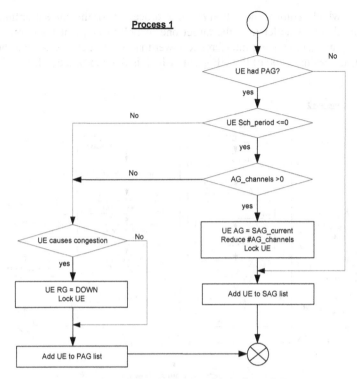

Figure 3-9: Splitting grants and handling CC inputs

Process 2: Interference and Scheduler Grants Handling

After the scheduler finishes the initial splitting of HSUPA users into primary and secondary categories, it starts executing process 2 which is shown in Figure 3-10. In this process, grants are allocated to UEs. In order to perform such allocation of grants for users, the scheduler maintains the target interference level of the cell. As mentioned at the beginning, the noise rise (NR) is the main resource and by calculating individual user specific load factors according to equation 3-3, these resources are quantified in the grants allocation process.

At the beginning of process 2 the current total uplink load factor is calculated using equation 3-1 and equation 3-2. To perform this calculation, PAG and SAG UEs are compared with their previous grants (previous data rates). The current uplink load factor is compared to the target uplink load factor which is calculated using equation 3-2. After this comparison there are two possibilities, either the current uplink load factor is greater than or equal to the target one. This means the current setup of the UEs and their grants are causing an interference above the target one, which requires the action taken in process-5, or the current uplink load factor can also be smaller than the target one, which means that the current UEs with their grants are causing interference lower than the target one, so that there is room for increasing some of the UE grants which is done in process 4.

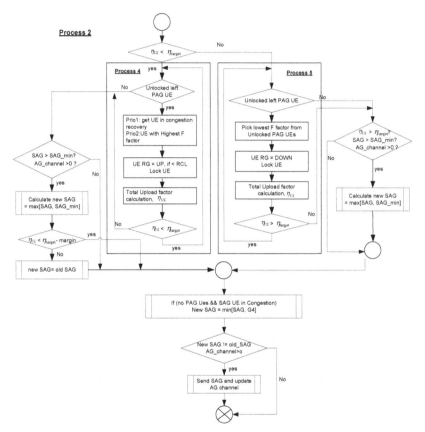

Figure 3-10: The grant allocating process for the uplink UEs

Both process 4 and process 5 work only on the PAG UEs. In order to increase the UE grants, process 4 starts searching for an unlocked UE i.e. it has not been processed yet or has not been allocated grants during process 1. The grant allocation for an unlocked user is performed based on two priorities. First a UE who is in congestion recovery phase requires increasing the grants, and second a UE with the highest F factor does so. The latter is defined as the relation between the RLC buffer occupancy and the UE data rate. It is a mechanism to offer fairness for the users who have large buffer occupancies with lowest data rates during previous TTIs.

When a UE is chosen, it is given a RG = "up" to increase its grant by one step of E-TFCI which provides the transport block size set information [61]. With the new UE grant assignment, the total uplink load factor is calculated and compared with the target load factor. In each turn, by giving grants for a new unlocked UE, this recursive procedure of load factor calculation and comparison is performed until it meets the target load factor requirement which represents the optimum cell capacity for the current TTI. If there is no unlocked UE to be given the grants, then the recursive procedure exits without meeting the requirement of the target load factor. In that case some part of the left resources can be used to increase the SAG grants of SAG UEs if only multilevel SAGs are enabled and supported by the scheduler.

The support of multilevel SAG is a configurable parameter for this scheduler. The remaining portion of the left resources is used to promote the SAG UEs to the primary mode by allocating the PAG grants. Setting up resource allocation for the above two categories is dependent on the vendor specific system requirements of the scheduler design. For example, the scheduler discussed here uses 30% of the left resources for the promoting process from SAG to PAG.

Handling Multi-Level SAG in Congestion

Handling multi-level SAG in case of transport congestion is shown in process 5 of the flow chart in Figure 3-10. It is similar to the procedure of process 4 but works in the opposite way. The scheduler starts executing this process when the current total load factor is higher than the target one. In this case, in order to reduce the total load factor, the scheduler decreases UEs grants by issuing RG "down" commands for some of the UEs. In this process, the UEs are prioritized based on the lowest F-factor which is the opposite action to process 4. It might also be the case when the scheduler exits process 5, and the total load factor is still above the target value. If the multi level SAG is enabled, the scheduler decreases the SAG by one step to reduce the current total load factor.

At the end of process 2, the scheduler checks the occurrence of a congestion indication from SAG UEs. The scheduler also reacts on such indications if there is no PAG UE in the system. Then new grants for SAG users are calculated as the minimum value between the current SAG and G4. The latter is the minimum reference congestion limit (RCL) between all SAG UEs causing congestion. The RCL value is provided by the transport congestion control mechanism and is discussed in chapter 4.

Process-3: Handling Promotion Of a SAG to a PAG UE

By finishing the process 2 functionality, the scheduler enters the grant promoting phase, which is represented by process 3 as shown in Figure 3-10. To promote one of the SAG UEs into the primary mode, there are several conditions that must be fulfilled as listed below.

- At least one AG channel should be available in order to provide the new grant.
- At least one SAG UE should be in the SAG UEs list.
- The total number of PAG should not exceed a predefined upper limit.
- The left resources should be sufficient to provide primary grants.

At the promoting process, the load factor is calculated by taking the target value into consideration. For this calculation, a factor K is defined as the difference between the maximum possible numbers of PAG UEs and the currently available number of PAG UEs. Therefore left resources are divided by this value K in order to provide fair resource allocation when promoting users. In addition to the fair resource allocation, priorities are also considered among eligible users based on their current buffer occupancies. This is maintained by another factor to consider fairness when it comes to delay constraints. By using the above consideration the first promoting user from SAG to PAG is selected and granted resources for the primary mode.

After scheduling grants allocation procedures, all SAG and PAG UEs transmit the data with their allowed grants in this TTI and the scheduler moves to the next TTI processing. After elapsing the TTI period, the whole scheduling process is again started from the beginning and continued until the end of the simulation.

3.4.2   Modeling HARQ

The modeling approaches of HARQ for HSDPA and HSUPA are similar in general. However, the HARQ mechanism for HSDPA is not modeled within the system simulator but modeled in the radio simulator. The traces of the per-user data rates (in terms of number of MAC-ds) were taken from the radio simulator after the outcome of the HARQ functionality. Therefore, when deploying traces for the downlink, effects of HARQ for the transmission over the radio channel were already considered.

For the uplink scheduler, no traces were used, therefore the HARQ functionality has to be modeled in the HSUPA network simulator according to the 3GPP specification [60, 62]. A simplified approach of HARQ functionality was taken into consideration for this implementation; the main functional features are listed below.

- Each UE has 4 HARQ processes and works based on a stop-and-wait protocol.
- Retransmissions of MAC-e PDUs are performed based on the air interface Block Error Rate (BLER).
- The following table shows the selected BLER with the number of retransmissions. It uses the concept of incremental redundancy and the following retransmissions have a higher recovery and hence a lower BLER for second and third retransmissions.

Table 3-1: HARQ retransmission probabilities

	1st transmission	2nd transmission	3rd transmission
BLER	10 %	1 %	0 %

The simplified HARQ procedure can be summarized as follows. When the MACe packet is transmitted, a copy of the packet is inserted into the HARQ process. Based on the BLER, for each packet, it is determined whether the transmission is a successful or not. If the latter is the case (assuming that a NACK is received), the same packet is retransmitted after elapsing of an RTT period. If the packet is transmitted successfully (assuming an ACK is received), the packet is removed from the HARQ process after the RTT period has elapsed. The RTT is set to "n" number of TTIs. For the current implementation, n is set to 4 based on allowed vender specific maximum delay limits. When a UE has a scheduling opportunity, it prioritizes retransmissions over new transmissions.

3.4.3 Modeling Soft Handovers

A soft Handover (SHO) has some negative influences on the network performance. Especially, the influence of an SHO on the network performance and the individual end user performance are significant at overload situations. A SHO leads to additional load inside the transport network. Therefore it causes blocking of certain parts of the BW for these additional overheads. The transport congestion control algorithm also has to handle such HO situations effectively. Therefore within the focus of the TNL feature development in the network simulator, it is important to consider the effect of the SHO load on the transport network. This leads to modeling of SHO effects in the uplink simulator.

For uplink transmission, the E-DCH scheduler has to control the offered load to the transport network. In a congestion situation, the scheduler should reduce the offered load to the transport network for some greedy connection in the uplink. These actions are performed combined with the transport network congestion control algorithm. Since SHO causes additional load over the transport network, the E-DCH scheduler has to manage the uplink transport resources among user connections during the soft handovers. However, a SHO is not critical for the downlink since the Node-B centrally manages the resources among connections and the flow control scheme along with the downlink transport congestion control scheme decides the offered load to the transport network. An overload situation is completely controlled by the DL congestion control schemes and further, in this case, the main buffering for the connections is located at the RNC entity. Therefore, for transport analysis, DL Soft handover does not play a significant role on the network and end user performance. Within the discussion of this section mainly the effect of the soft handover for uplink transmission is modeled.

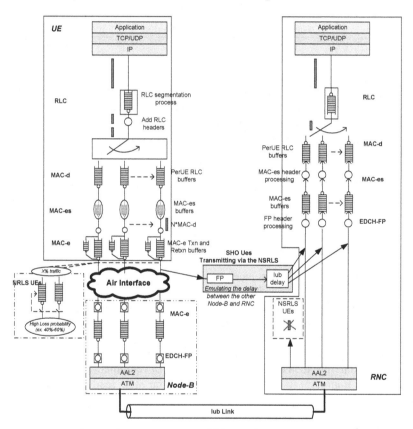

Figure 3-11: UL UE modeling with SHO functionality

For uplink transmission, two different SHO cases can be distinguished: the first case represents the soft handovers that some of the Serving Radio Link Set (S-RLS) UEs experience, in which those UEs will have their packets carried over the N-SRLS Iub link to the RNC. The second case represents the UEs from the neighboring cells that are in soft handover with the current cell which belongs to the considered Node-B. Those UEs are referred to as N-SRLS UEs, and some of their packets have to be carried by its Iub link.

Figure 3-11 shows the architectural overview of the UL UE modeling with SHO functionality. Further, it is shown that the SHO UE data is transmitted via the neighbouring Iub link and the N-SRLS UE data is transmitted over the considered Iub link. The first case (S-RLS UEs in SHO) is modeled in the following way:

a) The number of SHO UEs is chosen as a percentage of the total number of the HSUPA UEs.

b) The SHO UEs are considered to be in handover during the whole simulation time. This is done in order to emulate the additional load on the transport network.

When SHO UEs are transmitting a MACe PDU over the air interface, a copy of the packet is also transmitted via the Iub of the neighboring cells. Since these neighboring cells are not modeled in this simulator however, they are modeled in following way to counter the impact of the load offered.

A direct link was created in the model that links the HSUPA layer of the Node-B with the corresponding layer of the RNC. This link is used to emulate the effects of the N-SRLS Iub links, where the packets are transmitted over this link directly to the RNC by emulating a certain packet error rate (for example 40% – 60%) and a certain delay variation (for example, 10 ms – 40 ms). Therefore SHO UEs get the duplicate arrivals at RNC and they are filtered at the MAC-es layer in RNC.

The N-SRLS UEs that are in soft handover with the cell in consideration are modeled in the simulator as well. Since the N-SRLS UEs are not the main concern of this analysis, and also the per-UE evaluation is not intended for them there is no need of modeling the full protocol architecture for those N-SRLS UEs to generate their traffic, instead a simplified approach is used to generate the traffic for those UEs according to the following guide lines. The N-SRLS UEs are needed to block some of the Iub link capacity with their traffic to provide the realistic scenarios for the simulations, so those UEs should only generate MACe PDUs to be carried over the Iub (after being encapsulated as FP PDUs) and then discarded at the RNC when they are received.

3.5   TNL Protocol Developments

The network between the RNC and the Node-B is considered as the transport network and also as the Iub interface for the UMTS based UTRAN network. As mentioned in section 2.3, it has been initially designed with a high speed ATM based transport network, later using other technologies such as IP over DSL.

Supporting QoS guarantees for different services at minimum cost over the transport network is the main challenge for many broadband wireless network operators. With the development of high speed core network technologies in recent years, the delay over the transport network significantly reduced and created many opportunities to use packet based transport networks with service differentiation in order to meet the end user service guarantees.

In the focus of HSPA traffic (HSDPA and HSUPA best effort services) the end user and network performance are investigated using ATM based and DSL based transport network technologies. These two transport technologies are modeled in the HSPA network simulator for the uplink and the downlink. Next, details about the implementation of these two transport technologies are presented separately.

3.5.1   ATM based Transport Network

Rel'99 UMTS based real time traffic requires strict delay requirements whereas the HSPA based best effort traffic is more concerned on data integrity in comparison to strict delay constraints. ATM is one of the key technologies which use the cell based packet transmission in high speed wired network to guarantee the required QoS for different services. Since the cells are relatively small, this leads to a significantly higher overhead compared to the IP based data transmission. This effect is significant for the transmission of the best effort traffic over the wired network compared to the real time traffic which has often small packets in nature. For example in Rel'99 UMTS traffic, mainly the voice traffic has small data packets which can be effectively multiplexed within a cell and can achieve the best network performance due to a very high statistical multiplexing gain.

3.5.1.1   ATM Traffic Separation

ATM traffic separation is the key technique which is achieved through virtual paths (VPs) and Virtual Circuits (VCs). When cells are transmitted over the ATM network, connections and routing are set up based on virtual path identifiers (VPIs) and virtual channel identifiers (VCIs). Several traffic separation approaches have been investigated for the traffic over the transport network in order to investigate end user performance by evaluating packet loss ratio, delay, delay variation and throughputs [14]. Selecting a proper traffic separation is completely vendor specific and can be chosen based on service requirements. For the HSPA simulation analysis, a 3 VP based traffic separation approach is deployed and the basic architecture is shown in the figure 3-12.

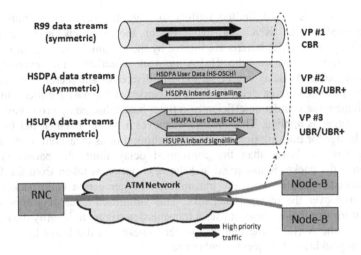

Figure 3-12: 3VP based traffic separation approach

The three VPs are used for the traffic separation based on their QoS requirements. VP1 is based on the CBR category and carries the Rel'99 UMTS based traffic, VP2 and VP3 are based on UBR+ and carry the HSDPA resp. HSUPA best effort traffic. Mostly Rel'99 real time services are symmetric services which require equal capacity in both directions with strict delay requirements. Therefore such services are mapped into the CBR VP with a guaranteed peak cell rate. In contrast to that, best effort services are asymmetric services which have a high demand in one direction in comparison to the other direction. For example, HSDPA mainly uses the downlink to transmit data and therefore requires a higher BW in the downlink compared to the uplink. The latter is mostly used to transmit the signaling information which is required by the downlink data flow. However, it is necessary to allocate a certain minimum guarantee for these signaling messages (mainly for in-band signaling). Therefore, the VPs which are assigned for HSDPA and HSUPA services are deployed using the UBR+ ATM service category.

3.5.1.2 AAL2 Buffer Management

AAL2 is used as the layer 2 adaptation layer for both Rel'99 and HSPA traffic services. The RNC side of the Iub interface consists of an RNC ATM switch which connects the Rel'99 traffic flows to the ATM multiplexer. The task of this intermediate MUX is to combine CBR and UBR PVCs at the RNC side of the transport network. This ATM MUX is designed together with a WRR (Weighted Round Robin) scheduler that guarantees specific service rates with the appropriate traffic configuration parameters.

AAL2 uses class based limited buffers to store the data. These buffers are implemented in the RNC and Node-B for the downlink and the uplink respectively. The AAL2 buffers are limited by the maximum delay limit which is a configurable value for the HSPA network simulator. For example, the maximum buffering delay can be set to 50 milliseconds and therefore all packets exceeding this maximum delay limit are discarded at the AAL2 buffers. In order to implement the simulation efficiently, a timer based discarding procedure with a packet trigger is applied. That means, whenever a packet is taken out from the AAL2 buffer for transmission, the packet waiting time is measured. If the packet waiting time is lower than the configured delay limit, the packet is sent. Otherwise the packet is discarded and the next packet is taken from the buffer. This procedure applies until the transmission trigger gets a valid packet to be transmitted over the transport network. This simplified discarding approach is efficient for simulations since it uses a transmission trigger to identify the delayed packets in the AAL2 buffer and also to send packets to the lower layer without having any additional frequent event triggers.

3.5.2   DSL Based Transport Network

The Digital Subscriber Line (DSL) technology is widely deployed for fast data transmission for industry, business and home usage. The great advantage of DSL is that it can use the existing telephone network by utilizing left frequencies above the voice telephony spectrum. This technology is one of the cheapest ways for fast data transmission which can be effectively deployed for data access with many other transport technologies such as IP and ATM.

The idea of the Mobile Network Operators (MNOs) is to use such a cheap fast transmission technology for the transport network in UTRAN inside the HSPA network (mainly for data access). This replaces the costly bundled E1 (ATM). Due to factors such as BER, delay and jitter which are also dependent on the line or cable distance, the performance over DSL based transport networks has not yet been investigated so far for HSPA networks. Therefore in this work, the performance of the DSL based transport network is investigated.

This section describes how the DSL based technology is implemented in the Iub interface between RNC and Node-B for data access in HSPA networks.

3.5.2.1   *Protocol Architecture for DSL based UTRAN*

As discussed above, DSL can be deployed combined with ATM or IP based transport technologies. Since IP over DSL is the most commonly used approach, it is implemented in UTRAN as a transport technology. The proposed protocol architecture is shown in Figure 3-13.

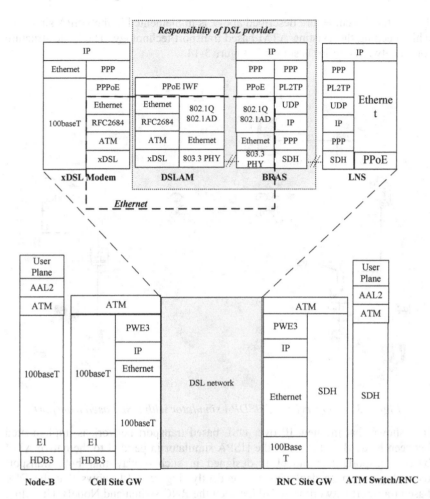

Figure 3-13: IP over DSL protocol architecture for the UTRAN

The above architecture shows that the Node-B is connected using cell site gateways (CSG) and the RNC is connected using RNC site gateways (RSG). The RSG and CSG use Pseudo Wire Encapsulation (PWE) in order to adapt the ATM layer to the IP layer. The DSL line is connected in between the xDSL modem and the Broadband Remote Access Server (BRAS). The traffic aggregation occurs at BRAS which connects the broadband network to the access network and vice versa. Further it performs the Point-to-Point Protocol (PPP) termination and tunneling. The Layer 2 Tunneling Protocol (L2TP) network server (LNS) which is located between BRAS and RSG provides the layer 2 tunneling.

The protocol architecture described above is implemented in the HSPA simulator while keeping the existing ATM based transport technology. The basic structure for all network entities is shown in Figure 3-14.

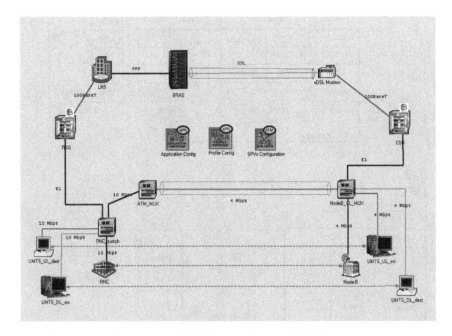

Figure 3-14: Overview of HSDPA simulator with DSL based transport

It is shown that the new IP over DSL based transport network is implemented between Node-B and RNC in the HSPA simulator in parallel to the native ATM based transport. The model is designed in such a way that both transport technologies can be deployed independently. Figure 3-14 shows that the DSL based transport network is added between the RNC switch and NodeB_UL_Mux. The RNC Site Gateway (RSG) is connected to the RNC via a switch. The LNS and BRAS are network entities in the service provider network; the DSL modem and BRAS are directly connected. Further, a DSLAM is included within the DSL network and the Cell Site Gateway is connected to the Node-B via NodeB_UL_Mux. In the model, the RNC switch and the NodeB_UL_Mux are the switchable devices which can turn ON and OFF the DSL based UTRAN based on the usage of ATM based technology.

The DSL protocol functionality is emulated between the xDSL modem and the BRAS by deploying a trace file which was taken from a field test. The trace file is applied at Ethernet packet level; it includes all behavioral aspects such as BER, delay and jitter of the DSL line including the DSLAM functionality. The main functionality of RSG and CSG is connecting two different transport technologies using the Pseudo Wire Emulation (PWE). The details about the generic network entities such as LNS, BRAS and DSL modem can be found in the respective product specifications, therefore further details will not be discussed here. This section focuses on further details about the implementation of DSL based UTRAN in the HSPA simulator.

3.5.2.2   CSG and RSG Protocol Implementation

The CSG and the RSG gateways are connected to Node-B and RNC respectively through the E1 based ATM links. To connect the ATM based network to the IP based Packet Switch Network (PSN), the Pseudo Wire Emulation Edge-to-Edge (PWE3) technology is proposed by the Internet Engineering Task Force (IETF). In this implementation, PWE3 is used to couple the two transport technologies, ATM and Ethernet via an IP backbone. The peer-to-peer protocols for the CSG and the RSG are shown in Figure 3-15.

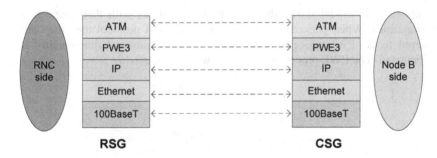

Figure 3-15: Peer-to-peer protocols in CSG and RSG entities

IP is the base media in between the two main transport technologies mentioned above. Both sides use a 100BaseT Ethernet link to connect to the other entities in the DSL network.

The basic PWE3 technology is described by the IETF standards. This protocol is implemented in the CSG and the RSG entities. The OPNET simulator uses the same node model for both entities and the required protocol functionalities are added to the process model of each node model. The node model is shown in Figure 3-16. The node model mainly consists of ATM, PWE IP and MAC (Ethernet) protocols. The IP and MAC protocols are used as transport for the PWE3 encapsulation. ATM is the base transport technology that is used by the UMTS/HSPA networks.

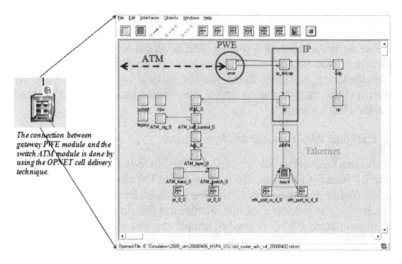

Figure 3-16: PWE, IP and MAC protocols in the CSG and the RSG entities

The above node model shows the main protocol entities with their respective functionalities. When packets are received by the ATM protocol then they are delivered to the PWE process model where all PWE protocol functionality is implemented. The PWE process model is shown in Figure 3-17.

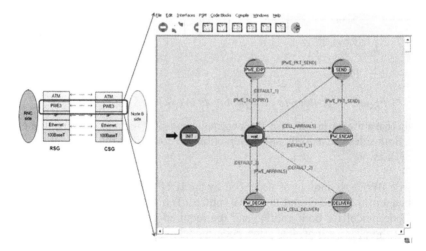

Figure 3-17: PWE protocol process model in the CSG and the RSG entities

The PWE packet is sent to the IP protocol which consists of two process models: the IP routing process model and the IP encapsulation process model. After

adding all functionalities (IP transport information) to the PWE packets at IP
level, the IP PDU is sent to the Ethernet protocol layer which consists of ARP and
MAC protocol processes. By adding all header information into IP packets, the
Ethernet frames are generated by the MAC protocol process and then the packet
is sent to the output port for delivery.

3.5.2.3   LNS, BRAS and xDSL Modem Implementation

The connections among LNS, BRAS and the xDSL modem are shown in Figure
3-18. The effect of the DSL link is modeled between the BRAS and the xDSL
modem by deploying a real trace file as it was discussed in the previous section.
In the OPNET simulator, all three network nodes use a single advanced node
model which is shown in the figure 3-18. All protocol functionalities such as PPP,
PPPoE, L2TP, UDP, etc. are implemented between the IP and MAC protocol
nodes in the advanced node model. The standard main IP process model is
modified and a child process which performs most of the protocol activities when
the connection is activated between respective nodes is added. The newly created
child process model along with all states is shown in Figure 3-19.

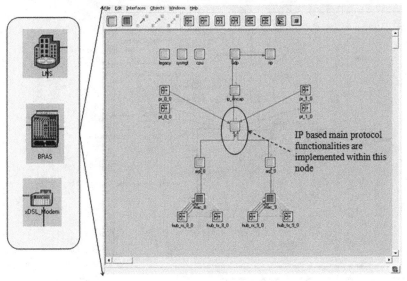

Figure 3-18: An advanced node model for LNS, BRAS and xDSL_modem

The protocol descriptions are coded inside the states of the child process. The
upper and lower states in the process model describe the packet arrival and
departure processes respectively, whereas the middle states describe the
processing and routing functionalities for the network entities. All other protocols

are coded inside each state appropriately. More details about the protocols and
their specifications are listed in Figure 3-20.

The PPP protocol uses two specification formats. When PPP is used over a direct
link, it uses a format according to IETF RFC 1662 [69] which specifies additional
fields such as address, control, and flags etc., compared to the PPP protocol
specified in the IETF specification RFC 1661 [68].

The L2TP protocol is implemented in the simulation model according to IETF
RFC 2661. This protocol runs over UDP; the OPNET default implementation is
deployed in the simulator. However, the newest version of L2TP call L2TPv3
defined by the IETF RFC 3931 [70] includes additional safety features compared
to the current RFC 2661 [71].

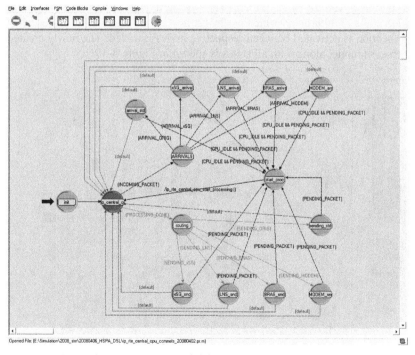

Figure 3-19: Process model for PPP, PPPoE, L2TP and UDP

The protocols which are newly implemented between BRAS and the xDSL
modem are listed in Figure 3-19. They are mainly PPP, PPPoE and the MAC
protocols. As in the previous implementation, IP and MAC use the default

OPNET implementation and the rest is implemented according to the specification given Figure 3-20.

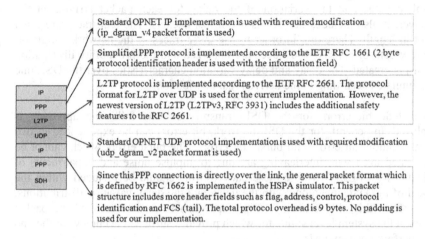

Figure 3-20: PPP, PPPoE, L2TP and UDP protocols specifications

The PPP is implemented according to IETF RFC 1661 [71] which was discussed in the previous section. The PPPoE protocol is implemented according to IETF RFC 2516 [72]. The header which causes an overhead of 6 bytes includes version/type, code, session_id and length fields.

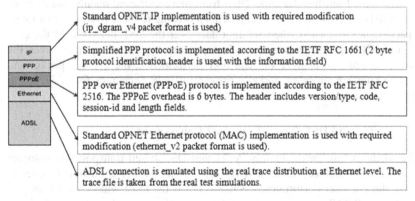

Figure 3-21: Added protocols between BRAS and xDSL model

3.5.2.4   DSL Trace Deployment

As discussed in the beginning of this report, the DSL line of the HSPA simulator is emulated using a real trace file taken at the Ethernet layer which records the absolute value and the variation of the delay for each packet arriving at the receiver. It also includes the details about lost packets. The real traces are often short in length. Therefore, in order to use these traces in the network simulator, the raw data of approx. 100 traces are analyzed and the probability distribution (CDF) is created. The delay and delay variation characteristics over the DSL line are simulated according to the derived real distribution.

To include bit errors for the DSL transmission, the following approach is deployed. In general, for the DSL network, bit errors can be experienced in two forms, random bit errors and bursty bit errors. The burst errors which cannot be eliminated even with interleaving are due to impulse noise and impairments of neighboring CPEs. Therefore these packet errors should be included within the model of the DSL line along with the delay and the delay variation distributions. To perform this, the following simple approach is used to emulate the packet errors in the system level simulator which provides sufficient accuracy for the end user performance analysis.

The main assumption of this method is that the total number of lost packets (or burst of lost packet events) is uniformly distributed over the given length of the trace file. First, all numbers of bit errors which create a corrupted packet and bursty errors which cause the loss of more than one packet are extracted from trace files based on their lengths. Then from these extracted details, the average packet and burst error rates are calculated. According to these average error rates, the packets are discarded when sending data over the DSL network. At the end, RLC or TCP has to recover those lost packets by resending them over the network. In order to analyze the end user performance, the average packet and burst loss rates are important and also provide sufficient accuracy for HSDPA and HSUPA networks.

3.5.3   ATM and DSL Based Transport Deployment

In the focus of this thesis, mainly ATM and DSL based transport technologies have been implemented, tested and validated in the HSPA simulator in order to analyze the end user performance. Since ATM is the most widely used transport technology, all TNL flow control and the congestion control analyzes which are described in chapter 5 and chapter 6 are performed using the ATM transport technology in the HSPA simulator. The DSL based HSPA analysis shows a similar performance to ATM based analyzes. Therefore, some of analyzes are given in Appendix A which mainly summarizes the impact of the DSL transport technology on the HSPA end user performance.

3.6   Radio Link Control Protocol

The radio link control (RLC) protocol is implemented between RNC and UE entities in the UTRAN. The Acknowledge Mode (AM) RLC protocol provides a reliable data transmission in the HSPA network for the best effort traffic. The packet losses over the Uu and the Iub interface need to be recovered using the RLC protocol and this avoids long retransmission delays by TCP which is located at the end entities. In this way, a minimum loss rate for the best effort traffic in the UTRAN network can be guaranteed and the required number of TCP retransmissions by the end entities is also minimized. Any retransmissions at the TCP level do not only cause additional delay but also consume additional network resources. Further, by hiding packet losses at the UTRAN network, TCP can provide effective flow control between end nodes to maximize the throughput and optimize the network utilization. In short, having RLC AM mode for best effort traffic within the UTRAN enhances the overall end-to-end performance. The following section briefly describes the RLC protocol implementation within the HSPA network simulator.

3.6.1   Overview of RLC Protocol

The RLC protocol does not only provide the AM mode operation but also two other modes: Transparent Mode (TM) and Unacknowledged Mode (UM). In TM mode, data will be sent to the lower layer without adding any header information. In UM mode, data will be sent to the lower layer with the header information, however similar to UDP, there is no recovery scheme. Hence it is recommended to use this mode of operation for most of the real time services which do not require strict constraints for the loss but which have critical limits for the delay. In AM mode, as described above, the RLC protocol works like TCP. In this mode of operation, a reliable data transmission is guaranteed between peer RLC entities by using an ARQ based recovery scheme. Further details about the RLC protocol are given in the 3GPP specification [22].

The RLC protocol is implemented in the HSPA network simulator. The main focus of this implementation is to protect the overall HSPA performance from packet losses occurring at the transport network for the best effort traffic. Packet losses occur either due to unreliable wireless channels or congestion caused by the transport network and they can be recovered by the RLC protocol without affecting the upper layer protocols such as TCP. The summary of the implemented protocol functionalities within the HSPA simulator is described in the next sections.

3.6.2   RLC AM Mode Implementation in HSPA Simulator

To achieve the objectives described above, some of the RLC AM functionalities specified in 3GPP [22] are implemented in the HSPA simulator. The implemented functionalities are marked with (✓) whereas others are marked with (✗) in the list given below. The RLC AM mode is applied to each user flow separately between RNC and UE.

✓ **Segmentation and reassembly:** this function performs segmentation and reassembly of variable-length upper layer PDUs into/from RLC PDUs.

✗ **Concatenation:** if the contents of an RLC SDU cannot be carried by one RLC PDU, the first segment of the next RLC SDU may be put into the RLC PDU in concatenation with the last segment of the previous RLC SDU. A PDU containing the last segment of a RLC SDU will be padded. The effectiveness of this functionality is not significant for a network which uses a large packet sizes such as best effort IP packets. Therefore, this function is not necessary for the HSPA model.

✓ **Padding:** If concatenation is not applicable and the remaining data to be transmitted does not fill an entire RLC PDU of given size, the remainder of the data field shall be filled with padding bits.

✓ **Transfer of user data:** this function is used for conveyance of data between users of RLC services.

✓ **Error correction:** this function provides error correction by retransmission (e.g. Selective Repeat) in acknowledged data transfer mode.

✓ **In-sequence delivery of upper layer PDUs and duplicate detection:** the data will be sent between peer RLC entities in a sequential order. The function detects any arrival of duplicated RLC PDUs and ensures in-sequence delivery to the upper layer.

✓ **Flow control:** this function allows an RLC receiver to control the rate at which the peer RLC transmitting entity may send.

✓ **Sequence number check:** this function is used in acknowledged mode. It guarantees the integrity of reassembled PDUs and provides a mechanism for the detection of corrupted RLC SDUs through checking the sequence number in RLC PDUs when they are reassembled into a RLC SDU. A corrupted RLC SDU will be discarded.

✓ **Protocol error detection and recovery:** this function detects and recovers errors in the AM operation of the RLC protocol.

✗ **Ciphering:** this function prevents unauthorized acquisition of data. Ciphering is performed in the RLC layer when the non-transparent RLC mode is used. In this simulation, ciphering is not used because the model is implemented in order to evaluate the RLC performance for the data traffic without considering the security aspects.

According to the RLC functionalities specified above, both RLC transmitter and the receiver entities are implemented in the UE and RNC nodes in the uplink and the downlink.

4   HSDPA Flow Control and Congestion Control

The flow control and congestion control are the key TNL features which have been developed to enhance the overall end-to-end performance for the downlink of the HSPA network. The motivation and theoretical aspects of flow control are described in this section. A new adaptive credit-based flow control algorithm is introduced for HSDPA. This enhanced flow control scheme has been implemented, tested and validated using the detailed HSPA simulator. Comprehensive simulation results and analysis of the flow control are presented in this chapter. Since the flow control scheme itself cannot avoid congestion in the transport network, a new congestion control concept has been introduced combined with the adaptive flow control scheme. First the concept of congestion control is described in detail and next, the different variants of congestion control algorithms are presented that have been implemented, tested and validated within the HSPA simulator for the downlink. Finally, a comprehensive simulation investigat
ion and analysis has been performed in order to analyze the end user performance. Further these flow control and congestion control schemes optimize the usage of transport network resources for the HSDPA network while providing aforementioned achievements.

HSUPA also uses the same congestion control scheme as HSDPA does. There is no requirement of implementing a flow control scheme in the uplink since the Node-B scheduler has full control over the transport and radio resources, and further there is no bottleneck behind the RNC in the uplink direction. However a congestion control scheme is required to avoid congestion at the transport level and to optimize the radio resource utilization. The congestion control input is taken into account in the implementation of the uplink scheduler as described in chapter 3.4.

4.1   HSDPA TNL Flow Control

Due to the bursty nature of the (HSDPA) packet-switched traffic, the accurate dimensioning of the TNL bandwidth is a difficult task. The mobile network operators (MNOs) are facing many challenges to optimize the performance of the TNL network to achieve best end user performance at minimum cost. Reducing the burstiness of the traffic over the TNL network and handling adequate buffer requirements at the Node-B buffers able to optimize the capacity requirement at the Iub interface [1, 2, 3, 4, 32, and 33]. When compared to Rel'99, HSDPA uses two buffering points at the Node-B and the RNC. To optimize the transport capacity, the data flow over the Iub interface should be handled effectively to cope with varying capacities at the radio interface [1, 24]. Thus, adequate queuing in the Node-B buffers is required. Due to large fluctuations of the time varying channel at the air interface, the Node-B buffers are frequently empty if less buffer capacity is used and such situations can cause wastage of the scarce radio resource [24, 25, and 32]. On the other hand, large buffer levels at the Node-B can result in large delays and also a large capacity requirement at the Node-B [26]. Further, a large amount of data buffering at Node-B leads to high packet losses during handovers resulting in poor mobility performance as well [24, 26, 27, and 28]. Therefore this is a clear trade-off which has to be handled carefully for the optimum network and the end user performance. As a solution to this issue, some of the literature discusses the need of flow control techniques [24, 25, 27, 29, and 34] that closely monitors the time varying radio channel demands and estimates Node-B buffer requirements in advance for each connection. However, the data flow handling over the transport network should be performed adaptively along with a feedback channel between RNC and Node-B. Therefore, an intelligent credit-based flow control mechanism has been introduced in this thesis which addresses all above requirements to achieve the best end-to-end performance at minimum cost while providing optimum network utilization.

The credit-based flow control scheme operates on a per flow basis and adapts the transport network flow rate to the time varying radio channel capacity. The estimated capacity requirements feedback to the RNC through the in-band signaling channels in an effective manner to minimized the signaling overhead over the uplink. The mechanism is further optimized to reduce the burstiness over the transport network that efficiently utilize the transport network providing minimum cost for the usage of network resources. By deploying different HSDPA traffic models, the overall network and end-to-end performance are analyzed using the HSDPA simulator which is discussed in chapter 3. During these analyzes and investigations, the performance of the developed credit-based flow control schemes is compared with the generic flow control scheme.

All the aforementioned aspects are described in this chapter in detail and the chapter is organized as follows. First, the generic flow control schemes (ON/OFF) and the adaptive credit-based scheme are discussed in detail and next, the performances of the two algorithms are presented and analyzed using the HSDPA network simulator. Finally, the conclusion is provided by summarizing the achievements at the end of the chapter.

4.1.1   HSDPA ON/OFF Flow Control

The ON/OFF flow control mechanism is considered to be the most simple mechanism which can be applied for the purpose of flow control for HSDPA traffic on the Iub link. The flow control mechanism monitors the filling levels of the MAC-hs buffers for each user flow. It uses an upper and a lower threshold to control the MAC-d flow rate over the Iub. The MAC-hs buffer with the two thresholds is shown in Figure 4-1.

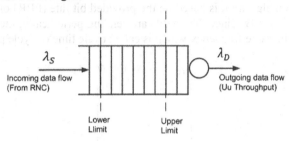

Figure 4-1: MAC buffer per flow basis

It is assumed that the incoming data rate to the MAC-hs buffer is λ_S (source rate) and the outgoing flow rate from the MAC-hs buffer is λ_D (drain rate). The upper and lower buffer limits are determined by an approximated value of the round trip time (RTT) between Node-B and RNC and by the Source/Drain rate as shown in *equation 4-1*.

$$th_{Upper} = [i-1] \cdot \lambda_S \cdot RTT$$

$$\text{where } i = 3, 4, ..., 10$$

$$th_{Lower} = \lambda_D \cdot RTT$$

equation 4-1

The per-user ON/OFF flow control mechanism assigns credits to each individual user in the Node-B based on the maximum achievable data rate over the air interface. If the upper limit of the buffer is exceeded, the data flow over the Iub link is stopped and if the buffer limit reaches the lower limit, data is transmitted over the Iub. This ON/OFF flow control mechanism protects the MAC-hs buffers from overflowing as well as emptying. However, when the incoming traffic has bursty nature, the delay variation significantly fluctuates from a lower to a higher value and vice versa. For such situations, the ON/OFF flow control mechanism turns out to be unstable [3], hence it becomes difficult to meet the requirements at the radio interface. On the other hand the ON/OFF flow control mechanism increases the burstiness of the traffic rather than mitigating it [4].

4.1.2   HSDPA Adaptive Credit-Based Flow Control

The adaptive credit-based flow control mechanism is developed to optimize the Iub utilization while providing required QoS to the end user in the HSDPA network. It smoothes down the HSDPA traffic and reduces the burstiness over the Iub interface. This flow control mechanism introduces two new aspects with respect to the ON/OFF flow control algorithm. First, the frame protocol capacity allocation message is sent periodically from the Node-B to the RNC. Second, the credit allocation algorithm is based on the provided bit rate (PBR) of the per-user queues in the Node-B. Since the credits are sending periodically, the PBR is also calculated at the same frequency which is called cycle time or cycle period.

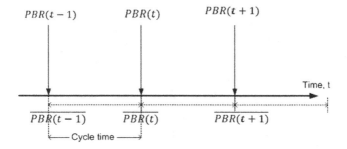

$PBR(t)$: Average number of MAC-ds at time t

$\overline{PBR(t)}$: Weighted Average value of $PBR(t)$

Figure 4-2: PBR calculation procedure

The HS-DSCH Provided Bit Rate (PBR) measurement is defined as follows in the 3GPP specification [22]: "For each priority class the MAC-hs entity measures the total number of MAC-d PDU bits whose transmission over the radio interface has been considered successful by MAC-hs in Node-B during the last measurement period (cycle time), divided by the duration of the measurement period ". Figure 4-2 shows the calculation of the PBR based credit allocation which is sent periodically after the cycle time. The PBR calculation is done at time t for the next allocation message sent after one cycle time considering the current and the previous successful transmissions. PBR(t) is the average number of MAC-d PDUs for each priority successfully sent over the air interface during a cycle time. Then, the average PBR rate, $\overline{\text{PBR}(t)}$ is calculated using the following formula with w being the weighting factor (for example default $w = 0.7$).

$$\overline{PBR(t)} = w \cdot \overline{PBR(t-1)} + (1-w) \cdot PBR(t)$$ *equation 4-2*

The filling level of the Node-B priority queue is calculated using the measured MAC-hs queue size qs(t) and the calculated PBR average according to the given formula below.

$$Filling \_ level = f(t) = \frac{qs(t)}{\overline{PBR}(t)} \quad (ms) \qquad \qquad equation\ 4\text{-}3$$

Figure 4-2 shows the overview of the buffer management for the credit-based flow control scheme. As in the ON/OFF flow control mechanism, there is an upper and a lower threshold that are maintained to control the flow over the Iub interface. Both the upper and the lower buffer limits are set according to the estimated value of the RTT. The credits to be sent over the Iub interface are calculated by using the MAC-hs filling levels and the upper/lower limit of the Node-B user priority queues. With the employment of the upper and lower queue thresholds, the filling level can be in three regions: less than the lower limit, higher than the upper limit or between these two thresholds. The number of credits allocated to the RNC for data transfer is calculated according to the three active regions of the MAC-hs buffer fillings and is given by the following three cases.

Case 1:
If filling level ≤ lower limit,
   then $Credits = 2 \cdot \overline{PBR}(t)$ *(MAC-d/TTI)*   *equation 4-4*

Case 2:
If lower limit ≤ filling level ≤ upper limit,
   then $Credits = factor \cdot \overline{PBR}(t)$ *(MAC-d/TTI)*   *equation 4-5*

Where the *factor* is a value between 0 and 2 which changes based on the variation of the filling level in MAC-hs buffer

Case 3:
If upper limit ≤ filling level,
   then Credits = 0 (MAC-d/TTI)   *equation 4-6*

The initial credit/interval calculation is performed using the Channel Quality Indicator (CQI) and is sent to the RNC as a Capacity Allocation (CA) message to regulate the data transfer over the Iub interface.

So the number of MAC-d PDUs transferred to each Node-B priority queue can be regulated in periodic instants of the cycle time according to their individual buffer occupancy levels. A user queue with less priority or a user with bad radio channel quality gets no MAC-d PDUs from the RNC, which otherwise would have been an unnecessary usage of the precious Iub bandwidth.

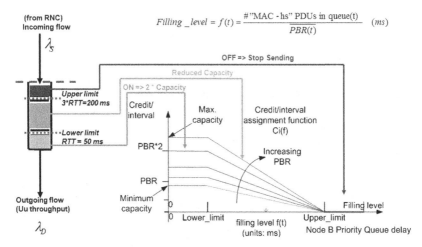

Figure 4-3: Credit-based buffer management technique

The credit-based flow control mechanism is implemented at FP and MAC-hs layers in the HSPA network simulator. Since flow control is only applied for the downlink data flow which can be centrally controlled by the node-B, the credit-based and generic ON/OFF flow mechanisms are implemented to control the downlink user flows. The credit-based buffer management is implemented in Node-B and the traffic regulation is implemented in the RNC layer. The FP is the main protocol which handles all flow control related functionalities.

The basic functional diagram of downlink flow control is shown in Figure 4-4. The flow data rate is determined using the calculated PBR and the filling level of the MAC-hs buffer in the Node-B. The credits/interval information or capacity allocations which are calculated according to the flow data rate at the Uu interface are sent to the RNC to calculate the frame size and the frame interval. The Node-B sends these HS-DSCH credits allocation messages periodically to the RNC in order to regulate the downlink data flow to the transport network. According to the allocated credits, the FP PDU is formed and sent to the Node-B by the RNC. All this flow control is performed on per user flow basis separately and independently.

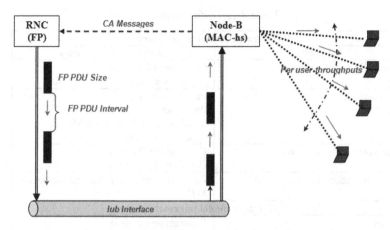

Figure 4-4: Credit-based flow control scheme functional overview

4.1.3   Traffic Models and Simulation Configurations

The most common type of HSDPA packet traffic consists of web and FTP traffic using TCP (Transmission Control Protocol) as a reliable end-to-end transport layer protocol. The web traffic model, defined by the ETSI standards [1 and 19], is selected to evaluate the performance of HSDPA traffic under moderate network load. The FTP traffic model is the worst case traffic scenario. It is used to overload the network in order to analyze the impact of the flow control features on the end user performance as well as the overall network performance. Under this traffic configuration, all users in the cell are downloading a large file and utilizing the network resources up to the maximum available capacity.

The parameters of the web traffic model defined by ETSI and the FTP traffic models [19 and 22] are given in Table 4-1. These traffic models are configured and deployed at the application layer of the end user entity so that all the other protocol effects can be included into the overall analysis. Especially, RLC and TCP protocols effects are investigated.

The simulation scenarios are defined based on two main analyzes to investigate the effect of the adaptive flow control algorithms on the performance of HSDPA network. Further investigations are performed to find the bandwidth recommendations for the Iub interface for these two different flow control schemes. Simulation analyzes provide detailed results to elaborate the effect of the adaptive flow control algorithm in comparison to the ON/OFF flow control algorithm.

Table 4-1: Traffic models

ETSI traffic model parameters	
Parameters	Distribution and values
Packet call interarrival time	Geometric distribution μ_{IAT} = 5 seconds
Packet call size	Pareto distribution Parameters: α=1.1, k=4.5 Kbytes, m=2 Mbytes μ_{MPS} = 25 Kbytes
Packet sizes within a packet call	Constant (MTU size) Constant (1500 bytes)
Number of packets in a packet call	Based on packet call size and the packet MTU
FTP traffic model parameters	
File size	Constant distribution μ_{MFS} = 12 Mbytes *(No reference, Only for testing purpose)*

As shown in Table 4-2, the simulation analysis 1 is performed based on the FTP traffic model whereas the simulation analysis 2 is done based on the ETSI traffic model which offers a moderate load compared the FTP traffic model.

Table 4-2: Simulation configuration

Simulation Analysis		
Analysis	**Simulation Configuration**	**TNL flow control**
Simulation Analysis -1 (using FTP)	Configuration-1	ON/OFF FC
	Configuration-2	Credit-based FC
Simulation Analysis -2 (using ETSI)	Configuration-3	ON/OFF FC
	Configuration-4	Credit-based FC

Each simulation analysis consists of two simulation configurations which are defined according to the selected flow control algorithm, either credit-based FC or ON/OFF FC. The simulation analysis 1 is used to study the effect of the two flow control mechanisms under the worst network load situation and whereas simulation, analysis-2 analyzes the above-mentioned effect under moderate HSDPA network load which is normally bursty in nature. For TCP, the widely used NewReno flavor is configured and the RLC protocol is set to run in acknowledge mode with its default settings in the HSDPA network simulator. Each cell is configured with 20 users being active in the cell until the end of the simulation time. The simulation model time is 2000 seconds which is sufficient to achieve steady state results for the analysis.

4.1.4   Simulation Analysis for FTP Traffic

As mentioned above, the HSDPA network simulator is configured with a very high offered load for this analysis. All users are downloading very large files simultaneously in the system utilizing full radio and network resources. Under these conditions, the effect of different flow control schemes is analyzed for the network and end user performance. The Iub ATM link, between RNC and Node-B is configured with 4 Mbit/s. Only one cell in the Node-B is configured with 20 UEs performing FTP downloads. The simulation results are analyzed from system level to application level.

4.1.4.1   *ATM Link Throughput*

The downlink ATM throughput between RNC and Node-B is shown in Figure 4-5. The actual variations of the link throughput for both FC schemes are elaborated along with model time.

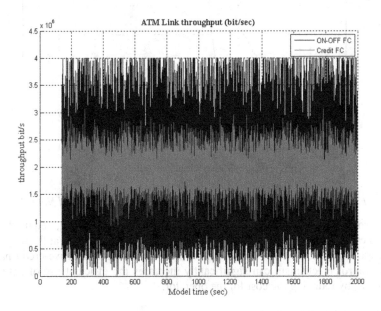

Figure 4-5: Iub ATM link throughput, between RNC and Node-B

Figure 4-5 depicts that the ON/OFF flow control shows a very high burstiness over the transport network whereas the adaptive credit-based flow scheme yields a low burstiness over the transport link. The adaptive credit-based flow control scheme effectively mitigates the burstiness over the transport network compared to the ON/OFF flow control scheme. This effect can be further seen from Figure 4-6 which shows the cumulative distribution of the ATM link throughput. The high burstiness leads to a high congestion probability at bandwidth limited networks. Further it requires high bandwidth and large buffers in the TNL network in order to meet the end user QoS.

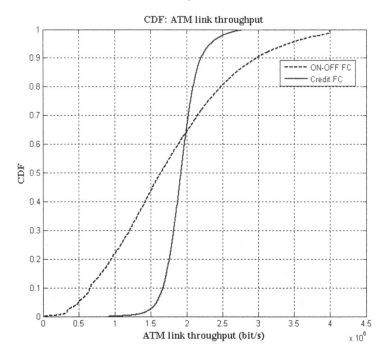

Figure 4-6: CDF of ATM link throughput between RNC and Node-B

By introducing the adaptive credit-based flow control algorithm, required BW can be even reduced less than 3Mbit/sec without any packet losses at the transport network compared to ON/OFF flow control. Therefore, the adaptive-credit based flow control mechanism can significantly increases the efficient use of network resources by reducing the cost for the operator.

4.1.4.2   *TNL Delay Performance*

The cumulative distribution of the end-to-end delay of the FP PDUs between RNC and Node-B for the downlink is shown in Figure 4-7. This statistic is measured from the point of time when an FP PDU is sent from the FP layer in the RNC to the point of time when it is received by the FP layer in the Node-B for the downlink data flow.

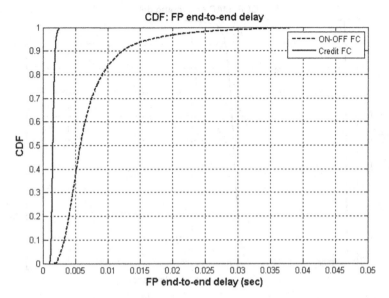

Figure 4-7: CDF of FP end-to-end delay

Figure 4-7 shows that adaptive credit-based flow control scheme provides a significantly lower FP end-to-end delay compared to ON/OFF flow control. Further the delay variation of the FP PDUs between RNC and Node-B is greatly reduced by providing excellent application performance.

4.1.4.3 MAC-d End-to-End Delay

Figure 4-8 shows the cumulative distribution of the MAC-d end-to-end delay for the adaptive credit-based and the ON/OFF flow control schemes. The credit-based flow control shows significantly lower MAC-d delay compared to the latter. The ON/OFF flow control mechanism has a very high buffer occupancy at the MAC-hs and ATM buffers which is due to high bursts at the TNL network. All these facts lead to a very high MAC-d end-to-end delay which includes all intermediate buffering delays as well. The credit-based flow mechanism has very efficient buffer management mechanism. It allows sending data from RNC based on the demand of the available channel capacity. Hence the MAC-d end-to-end delay is significantly reduced.

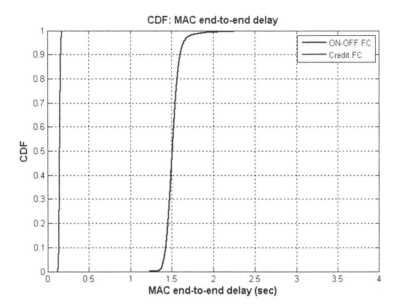

Figure 4-8: CDF of MAC-d end-to-end delay

4.1.4.4   *End-to-End Performance*

The overall IP throughput is shown in Figure 4-9. It is measured between end user entities and averaged over all users. It reflects the overall end-to-end performance of the UTRAN. The above figure shows that the achieved IP throughputs for both flow control mechanisms are similar and even slightly higher for the adaptive credit-based flow mechanism. That means both configurations achieve the same end user performance for different TNL conditions. Due to the reduced burstiness at the transport network, Figure 4-5 and Figure 4-6 show that the TNL bandwidth can be significantly reduced when applying the adaptive credit-based flow control mechanism compared to ON/OFF flow mechanism while achieving same end user performance which is shown in Figure 4-9.

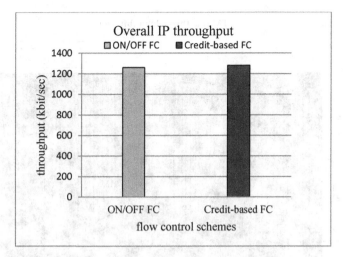

Figure 4-9: Overall IP throughput

4.1.5   Simulation Analysis for ETSI Traffic

This section presents simulation results for the simulation analysis 2 which deploys the moderate ETSI traffic model. The Node-B is connected to a cell with 20 users. Since the ETSI traffic model provides the moderate load to transport network compared to the worst FTP traffic model, the ATM link is configured to 3 Mbit/s bandwidth capacity for this analysis.

4.1.5.1   *ATM Link Throughput*

The downlink ATM throughput (in bit/s) is shown in Figure 4-10. As in the previous analysis, there is a clear reduction of the burstiness over the TNL network when the adaptive credit-based flow control mechanism is used. However, this traffic model is bursty by nature. Therefore the burstiness on the TNL network cannot be eliminated completely. Figure 4-11 shows the corresponding cumulative distribution function (CDF) of the ATM link throughput which exhibits the variations of the ATM throughput distribution.

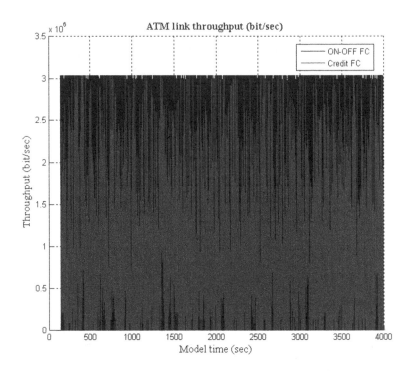

Figure 4-10: Iub ATM link throughput between RNC and Node-B

It shows a large number of zero occurrences in the throughput distribution. This means, the link is idle for a long period of time due to the frequent bursty fluctuations of the traffic. In comparison to the ON/OFF scheme, the adaptive credit-based scheme mitigates the idle periods over the transport network and also tries to optimize the link utilization by controlling traffic over the transport network.

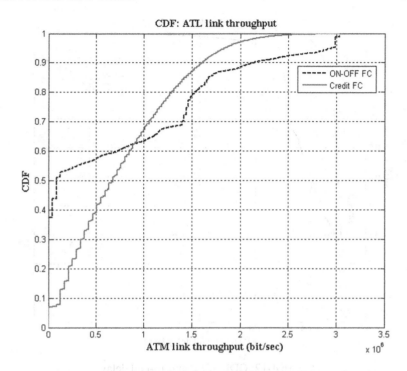

Figure 4-11: CDF of ATM link throughput between RNC and Node-B

4.1.5.2 TNL Delay Performance

The cumulative distribution of the end-to-end FP delay for the downlink is shown in Figure 4-12. It shows that the adaptive credit-based flow control configuration provides lower burstiness compared to the ON/OFF flow control algorithm. The latter also shows higher delay and higher delay variation compared to the credit-based flow algorithm.

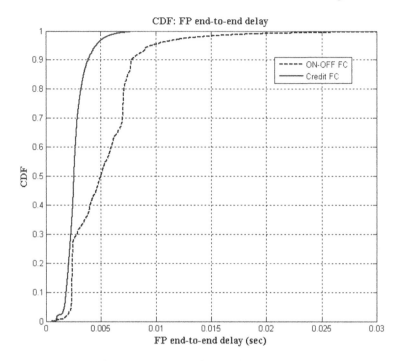

Figure 4-12: CDF of FP end-to-end delay

4.1.5.3   *MAC-D End-to-End Performance*

The cumulative distribution of the MAC-d end-to-end delay is shown in Figure 4-13 for the ON/OFF and the adaptive credit-based simulations. The latter flow control shows significantly lower delay compared to the ON/OFF flow control at MAC level which is measured between RNC and the UE entity. The adaptive flow control provides an enhanced MAC level delay performance.

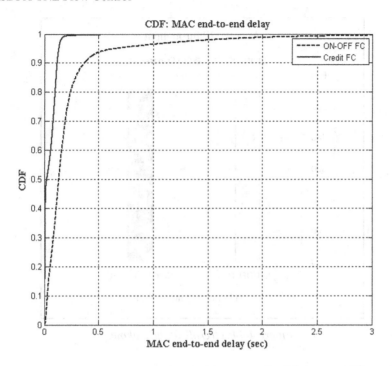

Figure 4-13: CDF of MAC-d end-to-end delay

4.1.5.4   *End User Performance*

The average IP throughputs for the ON/OFF flow control and the adaptive credit-based flow control are shown in Figure 4-14. This statistic shows the overall end user performance of the UTRAN. The achieved IP throughput is approximately equal for both simulations and even slightly higher for the adaptive credit-based flow mechanism. A lower burstiness of the carried traffic over the transport network was clearly shown in Figure 4-11 by the adaptive credit-based flow control mechanism compared to the ON/OFF flow control mechanism. Therefore, the adaptive credit-based flow control achieved the same application throughput with less congestion probability and also indicates that TNL has lower capacity requirement due to the reduced burstiness compared to ON/OFF flow control.

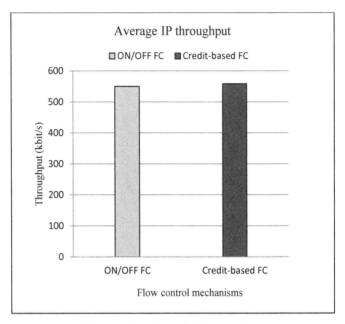

Figure 4-14: Overall IP throughput

4.1.6   TNL Bandwidth Recommendation

The TNL bandwidth recommendation is issued by deploying the ETSI traffic model which is the most moderate traffic for the HSDPA networks. The main quality of service criterion for the BW recommendation is 1% of the packet loss rate experienced at the TNL network. This means, that the maximum packet loss rate should be less than 1% at the bottleneck of the transport network. Two air interface scheduler types, Round Robin (RR) and channel-dependent have been deployed for the simulation analysis along with 3 cells attached to a single Node-B. In each cell there are 20 active users who provide offered traffic to the UTRAN network throughout the simulation time.

The selected simulation scenarios are listed in Table 4-3. The Round Robin scheduler is considered as the fair scheduler which distributes the fair share of the available capacity among the active users whereas the channel-dependent scheduler is an unfair scheduler which is used to optimize the throughput in the HSDPA network. The bandwidth recommendation is determined by varying the configured Iub ATM capacity and achieving the same QoS for two different flow control algorithms and applying two types of schedulers.

Table 4-3: Simulation configurations for BW recommendation

Traffic model	MAC-hs Scheduler type	#cell	Flow control type
ETSI based web traffic model	Round robin	1 cell	ON/OFF
			Credit-based
		2 cells	ON/OFF
			Credit-based
		3 cells	ON/OFF
			Credit-based
	Channel dependent	1 cell	ON/OFF
			Credit-based
		2 cells	ON/OFF
			Credit-based
		3 cells	ON/OFF
			Credit-based

Figure 4-15 shows the bandwidth recommendations when the round robin scheduler is used for three different cell configurations (1 cell, 2 cells and 3 cells) and two different flow control mechanisms. The required bandwidths are 2.9Mbit/s, 4.3Mbit/s and 5.7Mbit/s for one cell, two and three cell configurations respectively when ON/OFF flow control is used. These results also confirm the multiplexing gain of the ATM network. To add one additional cell with 20 users under this configuration requires approximately an additional half of the capacity being added to the previous cell capacity. This means approximately 50% of statistical multiplexing gain can be achieved for ON/OFF based simulations. The required bandwidths applying the credit-based flow control are 1.9 Mbit/s, 3.2 Mbit/s and 4.3 Mbit/s for one cell, two and three cell configurations respectively. These results show that by applying the adaptive credit-based flow control algorithm the required BW at the TNL network can be reduced about 33%.

Figure 4-16 shows the BW recommendations with the channel dependent scheduler used for three different cell configurations. The bandwidth requirements are 3.0 Mbit/s, 4.6 Mbit/s and 6.1Mbit/s for ON/OFF based simulation configuration for the one cell, two and three cell configurations respectively. The bandwidth recommendations when adaptive credit-based flow control is used are 2.2 Mbit/s, 3.9 Mbit/s and 5.2 Mbit/s for one cell, two cells and three cells configurations respectively. These results also confirm that by applying the adaptive credit-based flow control algorithm the TNL BW requirement at the TNL network can be significantly reduced.

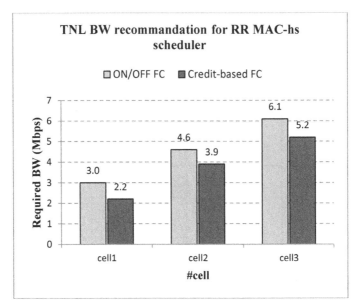

Figure 4-15: TNL BW recommendations for RR MAC scheduler

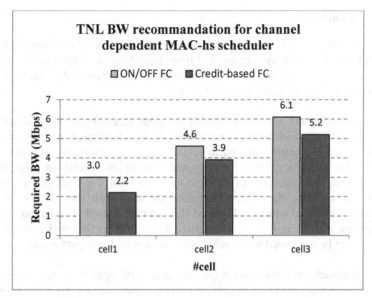

Figure 4-16: TNL BW recommendation for Max CI MAC scheduler

4.1.7   Conclusion

Within the flow control analysis, the theoretical and modeling aspects of the ON/OFF and the adaptive credit-based flow control mechanisms were discussed in detail. The performance of both mechanisms was presented, simulated and validated using an HSDPA network simulator.

When evaluating the simulation results for both simulation models which are based on the ETSI and the FTP traffic models, it is confirmed that the adaptive credit-based flow control mechanism achieves lower burstiness over the Iub link, lower buffer occupancy, lower FP delay and also the lower delay variation at the TNL network in comparison to the ON/OFF flow control mechanism. Based on above achievements, the required capacity for bandwidth recommendations is significantly reduced when credit-based flow control is used. The Iub bandwidth requirement and the utilization of resource at the transport network layer of the UTRAN can be optimized while achieving the same end user performance.

If the Iub capacity is limited in comparison to the cell capacity, congestion on the TNL network may occur. Congestion control techniques for the TNL network are addressed in the next section.

4.2   TNL Congestion Control

In the previous section, the effects of the flow control on the HSDPA performance were discussed in detail. Reducing the burstiness of the traffic over the transport network can greatly minimize the required transport capacity while providing required QoS for the end user. The flow control adapts the transport flow to the varying air interface channel capacity in such a way that it minimizes the burstiness over the transport network. However, the flow control mechanism does not consider the limitation at the last mile of the transport network. Therefore, congestion can occur at the transport network when it tries to send data at a rate which is higher than the available last mile capacity.

The required Iub TNL (Transport Network Layer) bandwidth for the HSDPA capable Node-B depends on the traffic engineering strategy pursued by the Network Operators (MNO), but one key aspect that MNOs always take into great consideration is that the "last mile" Iub connectivity is usually a major cost factor (e.g. large OPEX in case of leased lines). From all of the above, it follows that a typical TNL deployment for an HSDPA RAN foresees quite often transient situations in which the last mile Iub link is congested. Congestion at any level can cause severe performance degradation and major impact on the end user QoS requirements. For example, when congestion arises at transport level, a number of

RLC PDUs may be lost and the RLC protocol triggers retransmissions to recover those lost packets. Such retransmissions also introduce additional load to the transport network and can further enhance the congestion at the TNL network. In such situations, the end-to-end application throughput can be low if an appropriate Iub congestion control mechanism is not applied between the Node-B and the RNC. Therefore the HSDPA Iub congestion control feature is an important feature for PS (Packet Switched) RAB (Radio Access Bearer) using RLC acknowledged mode. According to 3GPP Rel-5 [22] the congestion control algorithm mainly consists of two parts: congestion detection (preventive and reactive) and congestion control. Two main congestion detection mechanisms implemented at the FP layer which means they are transport media independent, frame loss detection (reactive) by means of FSN (Frame Sequence Number) supervision and delay build up detection (preventive) by means of DRT (Delay Relative Time) supervision. In addition to these, a third congestion detection mechanism which is based on the checksum of the HSDPA data frames (reactive) is also considered. All these detection mechanisms work independently and produce indication triggers to the congestion control algorithm. This chapter introduces a complete solution for a suitable transport network layer congestion control mechanism which enhances the end user performance and Iub link utilization.

4.2.1 Overview of the HSDPA Congestion Control Schemes

The congestion control mechanism consists of two main functional entities as mentioned above: the congestion detection and the congestion control. The congestion detection identifies a congestion situation in the network before or while it occurs. The congestion control works based on the input of congestion detection triggers. Depending on the severity of the congestion, the congestion control module takes relevant measures to avoid congestion in the TNL network by regulating the output data rates. The granted data rates which are called "credits" are sent to the RNC as a capacity allocation message via the uplink in-band signaling channels. When the MAC-d flow is not in congestion, the granted credits are completely decided by the flow control mechanism; in this case, regular congestion avoidance (CA) messages are sent to RNC. In this HSDPA TNL implementation, the credit-based flow control, the congestion detection and the congestion control work together; they are shown in the functional block diagram of the Node-B in Figure 4-17.

Figure 4-17 shows that both congestion detection and congestion control algorithms are implemented in a cascade with the credit-based flow control mechanism which is effectively used to estimate the current air interface capacity in the Node-B for HSDPA. The congestion detection is triggered based on each arrival of an FP PDU at the Node-B. The figure also shows the two inputs to the congestion control algorithm: one is provided by the FC (Flow Control) module and the other by the congestion detection module. The FC and the congestion detection algorithms work independently and provide detection information to the congestion control module which processes all input information and triggers the capacity allocation messages to the RNC. The details about each algorithm which is shown in the block diagram are separately discussed in the next sections.

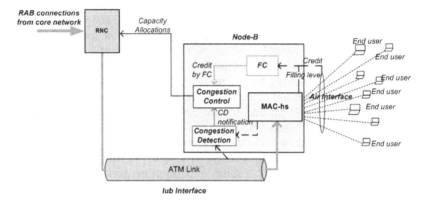

Figure 4-17: HSDPA FC and CC functional overview

4.2.2   Preventive and Reactive Congestion Control

The preventive congestion detection schemes can detect a congestion situation with high probability before it occurs in the network. Such detections can adapt the offered load according to the available transport capacity. Therefore, a prevailing congestion situation is mitigated and QoS is preserved. However, preventive congestion control schemes are not able to completely avoid congestion in the networks which experience bursty traffic. In such networks, in addition to the preventive congestion detection mechanism, the reactive congestion detection mechanisms are deployed to recover the congestion in the network. The combined effort of preventive and reactive mechanisms is also able to mitigate the congestion situation in the network. However, the effectiveness of all these schemes has to be tested and validated through proper investigations and analyzes which will be addressed in this chapter.

The preventive DRT based delay build-up algorithm monitors the FP PDU delay variation for the correctly received FP frames through the transport network for each MAC-d flow. High delay variation is often an indication for future congestion in the network. Figure 4-18 below shows the delay variation for several FP PDU transmissions with respect to the RNC and the Node-B reference counters.

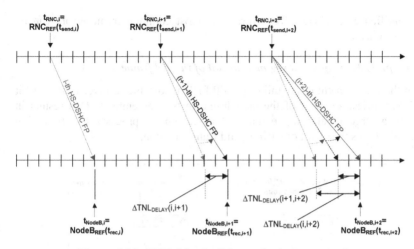

Figure 4-18: DRT delay build-up calculation procedures

According to Figure 4-18, the delay build-up algorithm monitors the TNL delay variations for each arrival of an FP frame at the RNC. For example, the delay variation between frame i and frame i+1 is given in the following formula.

$$\Delta(t_i) = \Delta TNL_{DELAY}(i, i+1), \quad i \geq 1 \qquad \textit{equation 4-7}$$

where $\Delta(t_i)$ is the delay variation between i and i+1.

The accumulated delay variation is calculated using these build-up delay variations and is given by $R(t_i)$,

$$R(t_i) = \sum_{i=2} \Delta(t_i) = \sum_{i=2} \Delta TNL_{DELAY}(i, i+1), \quad i \geq 1 \qquad \textit{equation 4-8}$$

Once the accumulated delay variation exceeds a certain user defined threshold within a certain time window, congestion detection is triggered to indicate foreseen congestion in the network.

Packet loss at transport level is a clear indication for a system overload. Therefore, the data flow over the transport network should be immediately controlled once a packet has been dropped. Such loss detection mechanism is called reactive congestion detection which is performed at the FP layer in the RNC of the HSDPA network. Since the TNL uses ATM as the transport technology between Node-B and RNC, ATM cells can be dropped when the system is overloaded. This can cause either generation of corrupted frames or loss of complete frames which is mostly due to bursty losses. When the header CRC check fails, the packet cannot be identified and is discarded. In many situations, the FP layer receives frames with a valid header but invalid payload. Such errors are identified as payload CRC errors. The most common frame errors can be categorized as follows.

Error type 1: Missing a last segment or tail of the FP frame

When the last segment or the tail of the FP PDU is lost, the receiver waits until it receives the last segment of the next frame to be reassembled. This results in creating a large invalid frame during the reassembly process. The following figure illustrates the effect of such invalid frame creation.

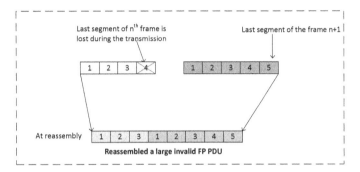

Figure 4-19: Invalid FP frame reassembly due to a missing tail of a frame

Error type 2: Missing segment except the tail of a frame

The receiver can reassemble a FP frame which is invalid due to a missing segment or due to an insertion of a foreign segment. Since the tail segment of the frame is preserved, the FP frame is reassembled by generating an invalid frame. A replacement of a foreign segment is a rare occurrence for the HSDPA network compared to a missing segment. Figure 4-20 illustrates such errors.

Error type 3: Loss of complete frame(s)

When a burst of cell losses occurs at the transport network, one FP frame or several FP frames can be lost. These kind of bursty losses are common for HSDPA PS networks when congestion persists for a long period of time.

Another important reason for CRC errors besides Iub congestion is PHY layer errors; therefore a mechanism for discriminating between rare PHY errors and real Iub congestion conditions is needed. This mechanism can be a probabilistic discriminator that discards PHY layer errors by taking into account typical BER figures of physical lines. This probabilistic discriminator is not implemented in the current HSDPA simulator for simplicity.

The above first two error types can be easily identified by the FP payload CRC check. The last error type can be identified by using a frame sequence number (FSN). All these loss indications are identified by the reactive congestion detection mechanisms which provide detection signals to the congestion control module. The next section discusses the congestion control process in general which acts based on these detection inputs.

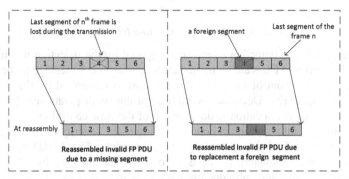

Figure 4-20: Invalid FP frame reassembly

4.2.3   Congestion Control Algorithms

The congestion control algorithm is triggered by the congestion detection events. A MAC-d flow mainly operates in two states: flow control and congestion control that are shown in the MAC-d flow state machine in the Figure 4-21. When there is no congestion in the network, the MAC-d flow is operating in flow control state. In this state the credits are updated periodically based on the air interface capacity changes. However, when congestion detection is triggered, credit calculation is completely handled by the congestion control algorithm which also takes the flow control output into account. The preventive congestion indications are considered to be light congestion indication inputs whereas the reactive congestion indications are considered to be severe congestion inputs by the congestion control module.

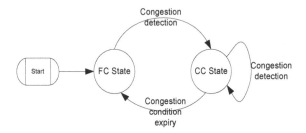

Figure 4-21: MAC-d flow state machine for congestion control

Once the congestion detection events are triggered by the detection module, the MAC-d flow goes to CC state by immediately sending a CA message to the RNC. At the congestion control state, the MAC-d flow is controlled by the Additive Increase Multiplicative Decrease (AIMD) algorithm with parameters (a, b) [26 and 28]. The credits allocation is done on top of the flow control credits whereas the amount of credit reduction is done according to the severity of the congestion detection trigger which is received by the CC module. The AIMD rate control mechanism with parameters (a, b) is used to reduce the rate by b% at each congestion detection and subsequently the rate is increased by "a" credits at the end of a user defined time interval in the recovery phase. The timer which is used to calculate the credits in congestion control state is also called congestion control step timer. Since the FC and CC algorithm work independently, the FC continuously monitors the air interface channel capacity and calculates the new credits at FC timer expiry, the timer which calculates the credits at the expiry in the flow control state. The congestion control module always compares both FC credits and CC credits, and only the minimum credit is sent to the RNC as a CA message. The condition for the MAC-d flow to exit the CC state is the value of credits calculated by AIMD (a, b) which either reaches again the value of credits when the MAC-d Flow entered the CC state or is exceeded by the flow control credits.

4.2.4   Congestion Control Schemes

Up to now, all principle aspects of the congestion detection and congestion control have been discussed in detail. In order to improve the HSPA performance by the best CC configuration along with the adaptive credit based FC scheme, several combinations of congestion detection and congestion control algorithms are selected and analyzed for different HSDPA traffic deployments. Mainly, three schemes which are defined depending on the functionality have been selected.

  (a) Reactive congestion control scheme (R_CC)

(b) Per MAC-d flow based reactive and preventive congestion control scheme (RP_CC)

(c) A common rate control based reactive and preventive congestion control scheme (RP_CC_Common)

The first two schemes (a) and (b) works per flow basis and are expected to be the appropriate CC schemes since they affects the individual flows which making congestion at the transport network due to overloading. On the other hand these schemes are designed to work fairly by distributing the available transport resources among the users. However, in order to further validate this expected behavior, the CC scheme (c) is selected for the analysis. All these investigation results are presented in chapter 4.2.6. Each of the above congestion control schemes and their functionalities are discussed in detail in the following sections.

4.2.4.1   *Reactive Congestion Control Scheme (R_CC)*

The FP payload errors are a common type of error in the HSDPA network. The first congestion control mechanism is built by deploying the FP payload CRC error based congestion detection mechanism and the congestion control mechanism. Both detection and control mechanisms work independently on a per MAC-d flow basis. This is a pure reactive congestion detection mechanism which detects congestion after it occurs. Since this is a severe congestion detection event, the congestion control module immediately activates the rate reduction procedures to avoid growing congestion in the network. As described in the previous section, the congestion control module uses the AIMD (a, b) control and recovery mechanism. In this scheme, FSN (Frame Sequence Number) based and delay variation based congestion detection mechanisms are ignored.

4.2.4.2   *Per MAC-D Flow Based Congestion Control Scheme (RP_CC)*

Since traffic has bursty nature, a complete frame or many frames together can be lost at any time. A high packet delay variation in the network is also a prior indication for such bursty losses. Therefore, a DRT based preventive congestion detection algorithm is used along with the reactive scheme. Furthermore, the reactive congestion detection scheme is also enhanced by adding the FSN based loss detection algorithm. Therefore, the reactive congestion detection scheme can detect both FP payload based CRC errors and bursty losses. All congestion detection triggers have been considered by the preventive and reactive algorithm during this implementation, therefore it is a comprehensive congestion control mechanism which is also called RP_CC in this context.

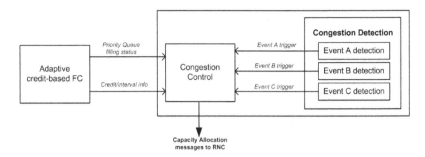

Figure 4-22: Reactive and preventive congestion control scheme

The reactive and preventive congestion detection algorithms and the congestion control algorithm work independently per MAC-d flow. Figure 4-22 shows the block diagram of the Node-B congestion control mechanism for a single MAC-d flow. All three types of congestion detection events are independently generated by the congestion detection modules. They are named as follows: event A is due to the detection of payload CRC errors, event B due to the detection of high delay variation and event C due to the detection of FSN based loss detection. Event A and event C are considered to be severe congestion indications whereas event B is not a severe congestion event. The congestion control activates the rate reduction procedures according to the AIMD (a, b) algorithm based on the detection of above events for that particular MAC-d flow. The amount of rate reduction is changed depending on the severity of the congestion which is indicated by the three detection events.

4.2.4.3 *Common Rate Control based Congestion Control Scheme (RP_CC_Common)*

Figure 4-23 shows an alternative congestion control algorithm for the RP_CC algorithm. The principle is similar since it uses the same congestion detection algorithms, however the way how the congestion control is applied and the ways how it reacts are different. Mainly, the rate reduction in the CC algorithm is applied on a per cell basis or per Node-B basis instead of a per MAC-d flow basis. When a congestion detection trigger is received by the CC module from one MAC-d flow, the congestion control activity is triggered for all MAC-d flows in the cell irrespective of the flow in which the congestion is detected. This means, when a MAC-d flow detects a congestion event, it immediately propagates this information to all other flows in the cell (or Node-B). Then congestion control is immediately applied to all MAC-d flows.

Figure 4-23 shows that there are "N" MAC-d flows in the cell. All detection events are connected in parallel to the congestion control modules of all MAC-d flows. Therefore, any congestion detection trigger or event (A, B or C) from one MAC-d flow is multicast to all congestion control modules and the rate reduction triggers are activated in parallel for all MAC-d flows as shown in Figure 4-23. Except for this behavior, all other functionalities are the same as in the RP_CC algorithm. For example, the congestion recovery and control activity are independently continued for each flow according to their current updated settings such as capacity limit, reference congestion limits etc.

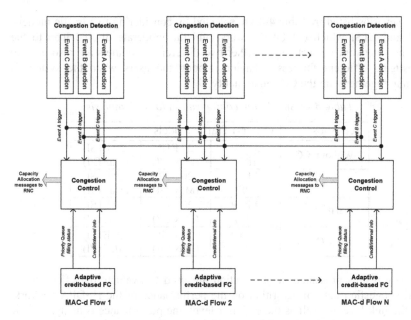

Figure 4-23: A common rate control based CC scheme

4.2.5   Traffic Models and Simulation Configurations

The HSDPA congestion control analysis also uses the same traffic models which were deployed in the HSDPA flow control analysis. The ETSI and the FTP traffic models which are given in chapter 4.1.3 are used for the simulation analysis. This section mainly highlights the simulation configurations which are used for the HSDPA congestion control analysis.

There are four scenarios used for the simulations. Three scenarios R_CC, RP_CC and RP_CC_common are defined according to the three congestion control schemes which were explained in the previous section; the remaining scenario does not deploy any congestion control scheme (without_CC). Further, all four scenarios are configured with 2 different traffic models (chapter 4.1.3). The FTP traffic model is used with two different user constellations and two different MAC-hs scheduler configurations: 3 FTP users/cell with the Proportional Fair (PF) scheduler at MAC-hs and 20 FTP users/cell with a channel dependent (MaxC/I) scheduler at MAC-hs. Further, the ETSI based web traffic model is configured with a 20 user constellation and the channel dependent scheduler at MAC-hs. All simulation scenarios which are defined are shown in Table 4-4 with their relevant configurations.

The scenarios listed in Table 4-4 are selected by considering two traffic models, FTP and ETSI which exhibit the worst and the moderate offered load to the transport network respectively. Further two different schedulers are also used to investigate different fairness behaviors among the users when the system is running together with the CC functionality.

Table 4-4: Simulation configuration for CC analysis

Scenarios	(#users) Traffic models		
Without_CC	3 FTP (PF)	20 FTP (Max C/I)	20 ETSI (Max C/I)
R_CC	3 FTP (PF)	20 FTP (Max C/I)	20 ETSI (Max C/I)
RP_CC	3 FTP (PF)	20 FTP (Max C/I)	20 ETSI (Max C/I)
RP_CC_Co	3 FTP (PF)	20 FTP (Max C/I)	20 ETSI (Max C/I)

The key results presented in this section are used to evaluate the effect of the TNL congestion control algorithm on the performance of the HSDPA network. For the link level as well as the end user level, the performance is analyzed. The ATM throughput statistics explain the utilization at the transport network whereas the IP throughput statistics show the performance at the end user entities. The delay statistics and the loss ratio statistics show the impact of the congestion control scheme on the TNL network.

4.2.6   Simulation Results and Analysis

In this section, all simulation results are presented and analyzed separately. All simulation scenarios are configured with the same parameters for all other protocols except the above mentioned changes. The ATM bandwidth is configured with 2Mbit/s for all simulation scenarios.

4.2.6.1 *ATM Link Utilization*

The average downlink ATM link utilization is shown in Figure 4-24 which illustrates link utilizations for the 3 FTP user scenario, the 20 FTP user scenario and the 20 ETSI user scenario under the effects of the three congestion control schemes mentioned before in comparison with no congestion control (without_CC). The link utilization is measured by dividing the average ATM link throughput by the configured link capacity which is shown in the following formula.

$$\text{Link Utilization} = \frac{\text{Average Throughput}}{\text{Configured Iub BW}} \qquad equation\ 4\text{-}9$$

As expected, the FTP traffic model based simulations exhibit a higher average link utilization than the ETSI traffic model based simulations.

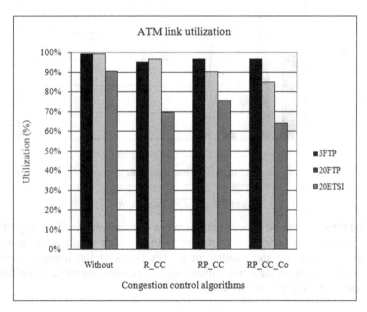

Figure 4-24: ATM link utilization

The link utilization figures for all CC algorithm based scenarios show less utilization compared to the without_CC scenario. Two arguments can be made on this indication. One is that high utilization is due to no control of the input traffic flow to the TNL network and the second is that the CC algorithms throttle down the input traffic unnecessarily causing less offered load to the TNL network. In order to clarify these issues further results at the higher layer protocols such as application layer throughputs are analyzed at the end of this section.

4.2.6.2   *Loss Ratio at TNL Network*

The loss ratio (LR) is measured in the AAL2 layer at the AAL2 buffers. The AAL2 discard mechanism is delay based which means the packets which have been waiting longer than 50 milliseconds in the AAL2 buffers are discarded. The LR at the TNL network is shown in Figure 4-25.

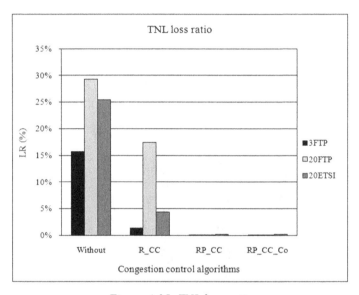

Figure 4-25: TNL loss ratio

From the figure, it can be noticed that there is a clear advantage of having RP_CC or RP_CC_common congestion control algorithms which significantly reduce the loss ratio by reducing the number of packet losses at the transport network whereas the simulation scenarios without any CC scheme experience significantly higher losses at the transport network.

The loss ratios under no congestion algorithm are 15.67%, 29.27% and 25.45% for the 3 FTP users scenario, 20 FTP users scenario and 20 ETSI users scenario respectively. The 20 FTP users scenario shows the highest loss ratio due to the worst case traffic model which offers heavy load to the network. Since the losses at the TNL network are significantly higher for without_CC simulation scenarios, the higher layer retransmissions such as RLC and TCP are consequently also high. These subsequent effects yield an increase of the offered load to the TNL network. This additional load is a wastage of network resources and prone to a congestion-induced collapse of the HSDPA network if it persists long in the system.

4.2.6.3   TNL Delay Performance

Figure 4-26 shows the average downlink FP PDU end-to-end delay for the three simulation scenarios.

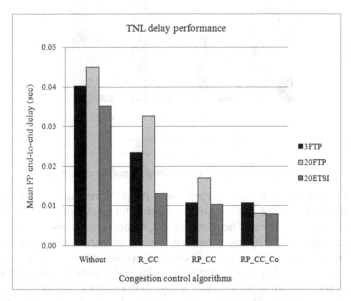

Figure 4-26: FP end-to-end delay performance

The highest FP end-to-end delay is shown for the without_CC simulation scenario whereas the lowest delay performance is given by the RP_CC and RP_CC_common algorithms. The R_CC algorithm which is based only on the FP payload CRC CC mechanism shows a moderate performance compared to other configurations.

4.2.6.4   Required In-Band Signaling Capacity over Uplink

Figure 4-27 shows the required uplink signaling capacity for each simulation configuration. The highest signaling capacity is required by the without_CC simulation scenario which completely operates under the credit-based flow control algorithm.

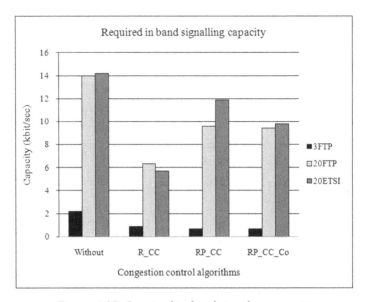

Figure 4-27: Required in-band signaling capacity

When the system is in flow control state the CA messages are being sent periodically based on the expiry of the FC timer which has a shorter duration than the AIMD recovery timer. Therefore these CA messages create the highest in-band signaling load on uplink compared to other scenarios as shown in Figure 4-27. However, when the system is in congestion control state, the number of CA messages which are sent over the uplink is reduced due to two main reasons. One is that the recovery timer period in the congestion control is higher than the flow control timer period. Therefore, the number of periodically generated CA messages is reduced when the flow is in CC state. On the other hand, the CA messages are only being updated if either the currently calculated credits are different from the previously sent credits or based on detection of new congestion events.

Since the FP payload based CRC CC algorithm detects the lowest number of CD events compared to the combined CC algorithm, it requires the lowest number of uplink in-band signaling bandwidth. The combined RP_CC_common algorithm shows a bit lower amount of signaling capacity compared to RP_CC algorithm. This is mainly due to fact that the congestion control activity for the RP_CC based configuration is only initiated for the MAC-d flow which detects the CD event while the other flows are still in the FC state.

4.2.6.5   *End User Performance*

The effective Iub utilization, shown in Figure 4-24 is the ratio between the effective application throughput which is measured at application layer and the theoretically achievable maximum application throughput with the configured Iub bandwidth.

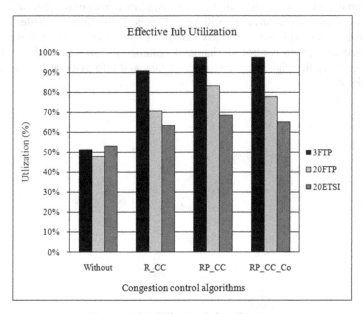

Figure 4-28: Effective Iub utilization

The theoretically achievable maximum application throughput can be calculated by taking into account the overheads of the RLC and ATM protocols. In this evaluation, the theoretically achievable maximum application throughput is estimated by multiplying the configured Iub capacity with a constant factor of 0.76 which is taken as a rough estimation of the overhead from ATM to application layer. From Figure 4-28, it can be seen that the highest effective Iub utilization is achieved by the RP_CC configuration for all simulation scenarios whereas the lowest effective utilization is achieved by the no congestion control based configuration for all simulation scenarios. The RP_CC_common scenarios show a slightly less effective utilization for the 20 FTP and 20 ETSI users constellation compared to the RP_CC based configuration. The FP payload based configuration shows less end user performance compared to the combined CC configurations but a significantly higher performance compared to the without_CC configuration.

4.2.7   Conclusion

The theoretical and modeling aspects of the preventive and reactive congestion control schemes have been discussed in detail within this chapter. There are several combinations of congestion control algorithms which have been widely investigated, tested and validated for the performance of the HSDPA network. The simulation results confirm that the combined usage of preventive and reactive congestion control algorithms achieves a significantly better overall performance for HSDPA networks compared to other configurations. Such combined congestion control mechanisms optimize the effective link utilization by minimizing the higher layer retransmissions and also achieve high end user throughputs.

5 Analytical Models for Flow Control and Congestion Control

Chapter 4 discusses the congestion control (CC) and the flow control (FC) algorithms in detail. Such algorithms are very complex in nature due to the dependency on many parameters. Therefore, analytical modeling of these algorithms has to be done carefully in a stepwise manner. During this investigation, first the behavior of the congestion control functionality is modeled and second, in a combined analytical model, flow control and congestion control are modeled together. Before getting into the detailed discussion of the analytical model, an overview of the two mechanisms is given.

The flow control and congestion control is done per MAC-d flow (user flow). The two states FC state and CC state are defined based on the flow activity whether the user flow is controlled by the FC algorithm or by the CC algorithm. The user flow stays in the flow control state when there is no congestion [3, 4] in the network.

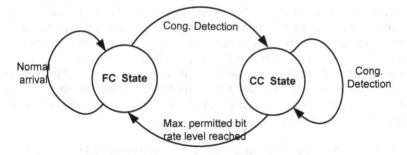

Figure 5-1: User flow state diagram

When the transport network is congested, the congestion control algorithm takes over the control of the data flow of the users and hence, user flows are rather staying in CC state than in the flow control state. The flow state diagram for a single user flow is shown in Figure 5-1.

The flow control algorithm estimates the capacity over the air interface and grants are provided to the RNC controller to send the data over the transport network on a per-flow basis. Even with the functionality of a proper flow control scheme, a user flow often gets into congestion at the transport network level due to limited transport capacity and bursty traffic.

The analysis of the transport network was performed using the simulations in chapter 5. However, it is a real challenge to investigate all the analyzes such as sensitivity analyzes and Optimization parameters for the CC and FC algorithms using the detailed system simulations which require a considerable long simulation time, often in the order of days, for each simulation run. Therefore, from this perspective it is necessary to have an analytical model which can perform these investigations and analyzes faster than the detailed system simulation does. Such an analytical model is addressed in this chapter which is structured into two main subchapters. Chapter 5.1 discusses the analytical modeling of congestion control, compares the analytical results and simulation results and finally provides a conclusion. Chapter 5.2 presents the combined analytical model of flow control and congestion control, compares the results with the simulation results and lastly gives a conclusion.

5.1   Analytical Model for Congestion Control

In today's Internet, many users who are connected via broadband access network consistently overload the system by downloading very large files. Such greedy users often create heavy congestion at the transport network level. The congestion control function in the HSDPA system controls such user flows in order to provide a fair share of the resources for all users. Due to the congestion control, most of the time, these flows are staying in the congestion control state and the CC algorithm adaptively allocates the grants based on the fair share of the available transport capacity. Investigation of such greedy user behavior at the transport level is the focus of this analysis.

The modeling approach of the congestion control is shown in Figure 5-2 for a single flow. The congestion control process which has an input and an output is shown. For the simplicity, this can be considered as a black box approach in which all CC functions are included. The input is described by the congestion indication (CI) interarrival process and the output can be described by granted data rate from the RNC according to the CC evaluation. Since all the processing for the CC is performed at the Node-B, the signaling transmission delay over the uplink is assumed to be constant.

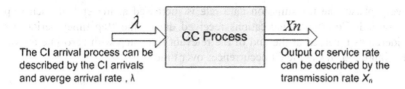

Figure 5-2: Block diagram of the CC modeling approach

The CI interarrival distribution depends on the current load at the transport network which again depends on several factors such as radio conditions, the number of users in the system and the type of user traffic. Modeling of the CI arrival process is discussed next.

5.1.1   Modeling of CI Arrival Process

The traffic burstiness in the transport network is mainly due to the HSDPA best effort traffic. All the user flows (MAC-d flows) are considered as independent and are modeled per flow basis. The behavior of the traffic over the transport network mainly depends on the bursty nature of the independent traffic flows and the time varying channel capacity at the air interface for each connection. Congestion often occurs randomly when several user flows which are operating at the same time try to transmit a bulk amount of data over the transport network simultaneously. This is a random process in which the current transmissions do not depend on the past actions but on the prevailing transport and radio channel conditions. Therefore, the congestion detection process for each flow can be assumed as a Poisson process in which the interarrival time is modeled using the negative exponential distribution in the continuous time domain.

The interarrival time between the arrival of the nth and the $(n+1)^{th}$ CI signals is denoted by a random variable T_n. The CI interarrival distribution $A_n(t)$ can then be defined as follows.

$$A_n(t) = \Pr[\, T_n \leq t\,] \qquad\qquad \textit{equation 5-1}$$

However, this analytical model considers the time in discrete intervals (or steps) since the data reduction procedures of the CC are activated in discrete time intervals as discussed in chapter 4. Therefore, all multiple occurrences within a single step are assumed to occur at the end of the time step and those multiple occurrences in a single discrete time step are also a Poisson distributed processes due to the memory less property of CI arrival process. When the system is in recovery phase, the transmission data rate is increased at the end of each step timer period. Congestion indications received during the step timer period are considered as received at the end of the relevant time step with zero interarrival time. Figure 5-3 shows the CI occurrences over time.

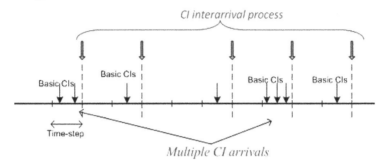

Figure 5-3: Modeling of CI interarrival process

First the CI interarrival distribution is modeled without considering multiple basic CI arrivals and then the basic CI arrivals are modeled separately. Since the multiple arrivals within a single step are a Poisson process, they can be modeled using the geometric distribution. The random variable K represents the number of CI arrivals in a single step. The probability of k CI arrivals in a single step can be represented by the following probability mass function.

$$\Pr(K = k) = (1 - p)^{k-1} p \qquad \textit{equation 5-2}$$

where p is the success probability of exactly one CI arrival within a step.

5.1.2   States and Steps Definition for the Analytical Modeling

Each CI arrival causes the current data rate to be reduced by a factor β which is equivalent to the parameter "b" of the AIMD algorithm which was described in chapter 4.2. When the MAC-d flow is in recovery phase and also further CI indications are not received then this indicates that the system can recover from the congestion. This means further data can be sent to the system to optimize the utilization. Therefore the data rate increases at a constant rate per AIMD step time which is called *step time* or *step* which is measured in seconds. At the end of each step, the data rate is increased by one step height which is "a" kbit/sec (discussed in chapter 4.2 as the AIMD parameter "a"). The terminology of steps and states is used throughout this analytical model. The CC functionality is shown in Figure 5-4, where the transmission rate is displayed versus the elapsed time. Further all CA messages are sent to the RNC through a dedicated uplink signaling channel, therefore the uplink in-band signaling (CA messages) delay is assumed as constant and it is also assumed that there are no packet losses during uplink signaling.

The state of the analytical model is defined as the transmission rate immediately after the CI arrival which means just after the CC decision has been taken. The discrete state which occurs due to CI arrivals is denoted by a random variable X.

$$X = \{ X_1, X_2, X_3, \ldots\ldots X_n \} \text{ for all n} \qquad \textit{equation 5-3}$$

The discrete state space or the set of possible values that X can take ranges from 0 to m, then m/α corresponds to the maximum possible step height that can be achieved by a single flow. $(m/\alpha) \cdot a$ [kbit/s] is the maximum possible data rate which is equal to the configured Iub capacity and $\alpha = 1-\beta$.

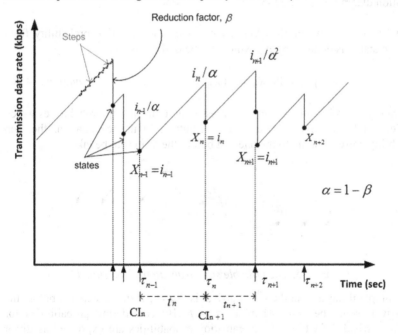

Figure 5-4: User flow behavior in CC state

As discussed above, the multiple CI events occurring within the same time step are modeled as simultaneous CI arrivals at the end of the corresponding step where the CC action is taken. Those multiple CI arrivals cause a multiple reduction of the data rate which is also called multiple drops. These multiple drops of the data rate are modeled as multiple departures in the analytical model.

For example, in Figure 5-4, there are two CI arrivals which are received by the CC process at the end of the step τ_{n-1} and the current rate is reduced by the factor β twice. If the state after processing CI is $X_{n+1} = i_{n+1}$, then the corresponding step before CI processing is given by i_{n+1}/α^2, where $\alpha = 1-\beta$. The complete behavior is described by the **_Discrete-time Markov Model with Multiple Departures_** and the next section describes this model in detail.

5.1.3   A Discrete-Time Markov Model with Multiple Departures

In this analytical model, the number of states mainly depends on the maximum network capacity and the step size. Assuming that the maximum number of possible states is m+1, then the random variable X can take state values from 0 to m. All possible state transitions from state $X_{n-1} = i$ are shown in the state transition diagram in Figure 5-5.

Since CI interarrival time has Markov property, the transition probability from state i to state j is denoted by $p_{i,j}$ which can be defined as follows.

$$p_{i,j} = \Pr\{X_n = j \mid X_{n-1} = i\} \qquad \text{\textit{equation 5-4}}$$

The transition probability p_{ij} can be expressed in terms of step probabilities which are derived from the CI interarrival distribution. In this calculation, the step probability corresponding to the one just before the drop is to be taken.

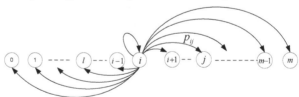

Figure 5-5: The possible state transitions from state i

For example if the state at the nth CI arrival is $X_n = j$ then the step just before the CI arrival would be "j/α" where $\alpha = 1- \beta$. By considering probabilities for multiple arrivals of CIs the final transition probabilities are expressed as given below.

$$p_{ij} = \sum_{k=1}^{d_{max}} \left[\Pr\left(\frac{j}{\alpha^k} - i\right) \cdot \Pr(k) \right] \qquad \text{\textit{equation 5-5}}$$

Pr(k) represents the probability of exactly k CI arrivals within a step and Pr(y) is the probability of the CI interarrival time with exactly y steps where $y = [(j/\alpha^k) - i]$ and d_{max} represents the maximum number of CI arrivals within a step.

The transition probability matrix is a square matrix of dimension $m+1$ and can be given as follows:

$$P = [p_{ij}]_{(m+1)x(m+1)} \qquad \text{equation 5-6}$$

In general, when the steady state transition matrix is known, the stationary state probabilities π can be calculated by using equation 5-7 along with the condition that sum of all steady state probabilities equals one. To solve this linear system equation, the recursive approach of sparse Cholesky factorization [80] is used.

$$\pi = \pi \cdot P \qquad \text{equation 5-7}$$

where π denotes the state vector, $[\pi_0, \pi_1, \pi_2, \pi_3, \ldots, \pi_m]$.

5.1.4 Average Transmission Rate

The performance is evaluated as average throughput or average transmission rate. The average throughput can be calculated by taking the area under the curve of the transmission rate divided by the total time (see Figure 5-6). The area under the curve gives the volume of the data which is transferred over the period of time. Normally, the time is assumed to be a very long period for the average calculations. Using the trapezoidal rule, the average throughput (volume under the curve / duration) can be calculated using the following formula.

$$E[X] = \sum_{k=1}^{d\,\text{max}} \left\{ \frac{[(i_1 + (i_2/\alpha^k)r_k)/2]t_1 + \ldots + [(i_n + (i_{n+1}/\alpha^k)r_k)/2]t_n + \ldots}{t_1 + t_2 + \ldots t_n + t_{n+1} \ldots} \right\}$$

$$E[X] = \frac{0.5}{\left[\sum_{j=1}^{\infty} t_n\right]} \left\{ [i_1 t_1 + i_2 t_2 + \ldots + i_n t_n + \ldots] + [i_2 t_1 + i_3 t_2 + \ldots + i_{n+1} t_n + \ldots] \sum_{k=1}^{d\,\text{max}} \frac{r_k}{\alpha^k} \right\}$$

$$E[X] = \frac{0.5}{\left[N\bar{t}\right]} \left\{ \sum_{i=1}^{m} i(N\pi_i \bar{t}) + \sum_{i=1}^{m} i(N\pi_i \bar{t}) \cdot \sum_{k=1}^{d\,\text{max}} \frac{r_k}{\alpha^k} \right\} = 0.5 \left[1 + \sum_{k=1}^{d\,\text{max}} \frac{r_k}{\alpha^k} \right] \sum_{i=1}^{m} i\pi_i$$

where $t_n = \tau_{n+1} - \tau_n$, $\bar{t}$ = Avg. CI interarrival time, r = drop prob., $\alpha = 1 - \beta$, d_{max} = max. num. of drops, N = total num. of trapezoids

The final average throughput calculation formula is as follows:

$$\text{Average Throughput} = 0.5\left[1 + \sum_{k=1}^{d_{max}} \frac{r_k}{\alpha^k}\right]\sum_{i=1}^{m} i\pi_i \qquad equation \ 5\text{-}8$$

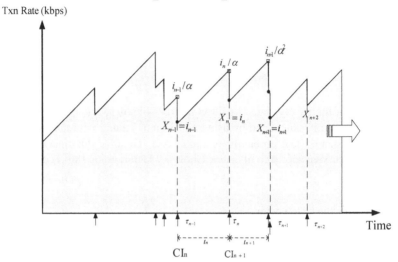

Figure 5-6: Flow behavior in CC

5.1.5   Simulation and Analytical Results

The simulation results and the analytical results are presented and compared in this section. First, all configurations of the input parameters are described in detail for both models and then the results for each model are discussed individually. Secondly, the analytical results are compared with the simulation results in order to validate the analytical approach for different configurations. Finally some of the sensitivity analyzes are done using both models and the performance is compared.

5.1.5.1   *Parameter Configuration*

The HSPA simulation model which was described in chapter 3 is used for the simulation analysis. The presented analytical model focuses on the congestion control functionality. In order to compare the performance between analytical model output and simulation output, the simulation model is also required to be configured in such a way that at least a single user flow is always in the congestion control phase all the time. The general user flow behavior is that when the system is congested, the user is in congestion control state, otherwise the user is in the flow control state. An off-period user does not transfer any data and is in neither congestion control nor flow control state. However, in order to compare the analytical model results with the simulation results, one user flow is required to be in CC state always. Some greedy flows may have a lot of data always to be transmitted and can be used for this analysis. The simulation scenario is configured in such a way that one UE flow always stays in the CC state for this analysis. The simulation scenario is configured including 20 users with two different traffic models. One user is configured with the FTP traffic model while all the remaining 19 users are configured with the ETSI traffic model [23]. Since the FTP user is downloading a large file he always stays in the congestion control phase until he finishes the download. In this application configuration, the FTP user is configured with very large files with zero interarrival time between two consecutive files so that the user always has data to be transmitted and stays in the CC phase.

The common congestion control parameters which were used for the analytical and the simulation model (in chapter 4) are listed below:
- AIMD CC step duration = 500 ms
- Step size of data rate changes = 33.6 kbps (1 RLC packet/10 ms)
- Congestion reduction factor β = 0.25
- Safe timer = 80 ms
- Link bandwidth = 2 Mbit/sec

In addition to the above, the simulation is configured with various other default parameters for other protocols such as TCP, RLC, and MAC etc. The simulation is run for about 4000 sec.

The analytical model assumes the geometrically distributed CI arrival process with a mean CI interarrival time of 0.2 CIs/step. The mean value is derived using a trace taken from the detailed system simulator. However, a fast queuing simulator which has been described in chapter 5.2.5 can be deployed to get this input parameter instead of using the detailed system simulation The fast queuing simulator provides faster results in comparison to the detailed system simulator which requires a long period of time in order to perform the same analysis. The step size is equal to the AIMD CC step duration, i. e. 500 ms as given in the configuration. Next, the simulation and analytical results are described separately and compared at the end of this section.

5.1.5.2   *Analytical Results*

First, all transition probabilities were calculated using equation 5-5. It is a square matrix of 37 states for the given scenario. Figure 5-7 shows the graphical overview of the transition probability matrix. The y-axis and x-axis show the state i and j respectively. Figure 5-7 depicts that when the flow is in a lower state, the transition probabilities to higher states are high. That means, there are more opportunities for jumping to the higher states from the lower states. However, when the flow is in higher states then there are more opportunities to jump to the lower states. This phenomenon completely agrees with the real behavior of the HSDPA system. This means, the system at higher states is more short-lived and jumps occur to lower states due to high congestion probability. When flows stay at higher states, more data is transmitted to the network which leads to a network overload situation. Therefore, there are more CI signals that cause rate reduction at the congestion control module in order to reduce the MAC-d flow data rate and to overcome congestion situations in the transport network.

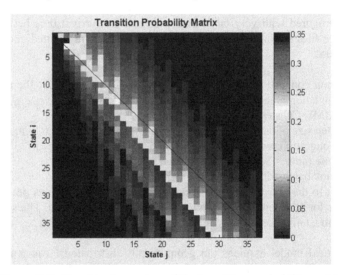

Figure 5-7: Graphical overview of the transition probability matrix

On the other hand, when the user data rate is reduced too much, the system is relaxed and the user is allowed to again send more data. Therefore, there are more opportunities to go for the higher states. Since the HSDPA traffic is bursty in nature and varies over time, the congestion control always tries to optimize the transport capacity. Further, the figure shows that a balance of transition between higher and lower states is achieved at the states between 12 and 18. At these states, the user transmits at data rates close to the average capacity which is allowed by the network share. However, when more capacity is available to transmit because the other users send less traffic for a particular period of time, then the flow rate is increased by the CC algorithm to utilize the unused capacity. The situation is reversed when all users have data to transmit over the network.

Therefore user flows always fluctuate according to the available transport capacity.

Figure 5-8 shows all stationary state probabilities from the analytical model that are calculated using *equation 5-7*. The figure illustrates that the middle states (12–18) have high state probabilities, indicating that the user is more often granted data at those transmission rates.

This indicates the average transmission rate which is allowed by the transport network, based on the shared capacity of the network. As described above, the user is allowed to transmit with higher and lower data rates based on the available current capacity of the network. These opportunities are quantified by lower and upper state probabilities.

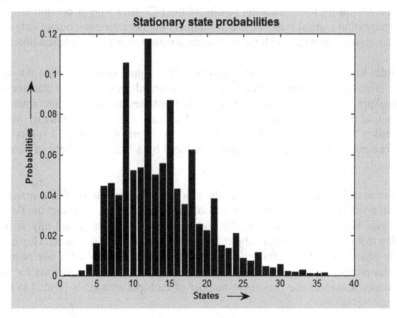

Figure 5-8: Stationary state probabilities

Further, the state probability shows a regular pattern of high state probabilities for some states compared to others. This is indicated by the high bars relative to the neighbor states in Figure 5-8, which is mainly due to the configured value of β set to 0.25 and multiple CI arrivals. The current state can change by a factor of 0.25 for each CI arrival. For example, if the current step is 20, then based on the arrival of a CI, it jumps down to 15 and based on two consecutive arrivals to 12 etc. For multiple departures, some regular states get a slightly higher probability compared to others which depends on the configured β value. During the simulations, it is observed that if the value of β changes, the pattern of the state probability peaks is also altered. However, these results are shown for the default β of 0.25. Later in this analysis more results are compared for different β values

Using equation 5-8 the average throughput for the analytical model is calculated. The average throughput is 552.52 kbps for the given scenario. Later these analytical results are compared with the simulation results in chapter 5.1.5.4 for both the stationary state probabilities and the average throughputs.

5.1.5.3   *Simulation Results Analysis*

The simulations were performed using the parameters specified in the previous section 5.1.5. One simulation run is selected to illustrate the results in detail. Figure 5-9 shows the simulation results of the FTP user who is continuously in the congestion control state. Figure 5-9 depicts the instantaneous output data rate or transmission data rate with its time average values over the simulation time.

In order to understand the exact behavior of the output data rate when the user flow is in the congestion control state, the zoomed view of the instantaneous throughput is shown in Figure 5-9. It clearly illustrates the stepwise increment of the output transmission rate between two CI arrivals. The CC functionality works according to the AIMD principle as explained in chapter 4. When a CI is received the transmission data rate of the user is reduced by a factor of β from its current value.

In one occasion, it is indicated that there are two CI arrivals (multiple arrivals) within a single step. The safe timer is configured with 80 ms to protect the flow's data rate being reduced too rapidly due to bursty CI arrivals, and any arrival during that time period is not subject to the transmission rate reduction. However, at the end of the 80 ms time period, one rate reduction is applied if one or more arrivals present during this time period. The step timer is about 500 ms for this simulation therefore the maximum number of multiple arrivals is limited to 6 in this consideration.

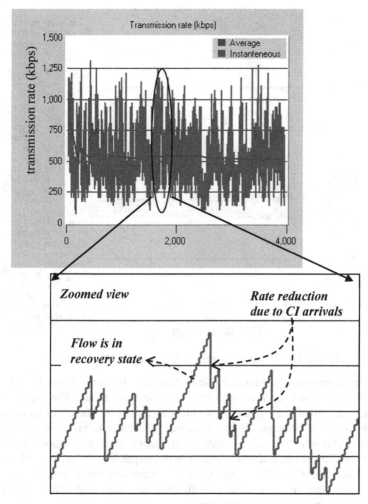

Figure 5-9: Instantaneous output data rate and its average value

To investigate the confidence level of the simulation results, 30 replications of the simulation were performed. Before explaining all the results, the next section elaborates the way that confidence intervals are calculated for replications. The confidence interval is calculated by taking the means of independent simulation replications. 30 simulation runs were performed with different random seeds for the 4000 sec model time. During the CI calculation, the warm-up period is not considered for the replications.

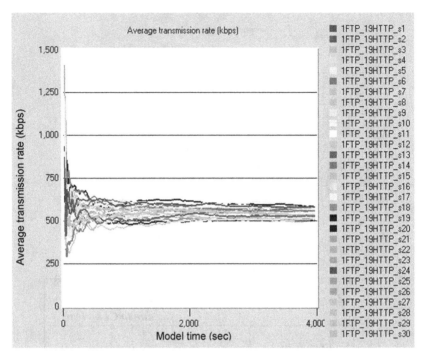

Figure 5-10: Average transmission rate for 30 replications

The instantaneous output data rate (simulation output) from 500 sec to 4000 sec was chosen for each replication to calculate the mean value. The mean of the replication is called the replication mean in this context. The replication means coincide with the principle of batch means. The latter is taken from a single simulation run whereas the replications are taken from simulation runs with different random seeds. Each individual run is independent; therefore replications lead to less correlation effects compared to batches of a single run. Figure 5-10 shows the replication mean values for all simulation runs and Table 5-1 depicts the replication means for 30 simulations.

Table 5-1: Replication means for 30 simulations.

Replications	1	2	3	4	5	6	7	8	9	10
Mean	592	524	539	518	576	587	555	531	565	523
Replications	11	12	13	14	15	16	17	18	19	20
Mean	540	538	574	537	512	572	602	514	506	610
Replications	21	22	23	24	25	26	27	28	29	30
Mean	561	507	579	533	566	540	515	575	484	538

All replications use different simulation seeds. Therefore, by construction, the replications are independent and are assumed as identically distributed which means that they are normally distributed with unknown mean and variance.

The point estimates such as the expected mean, variance and the confidence interval for all replications can be calculated as follows [78].

$$\text{Mean}, \bar{Y} = \frac{1}{n} \sum_{i=1}^{n} \bar{Y}_i = 547.07 \text{ kbit/s}$$

Where $\bar{Y}_i$ is the ith replication mean and n is the number of replications

$$\text{Variance}, V = \frac{1}{n-1} \sum_{i=1}^{n} (Y - \bar{Y}_i)^2 = 994.60 \text{ (kbit/s)}^2$$

The 95% of confidence interval, CfI

$$\text{CfI} = \bar{Y} \pm \left(t_{\alpha/2, n-1} \cdot \sqrt{V/n} \right) = 547.07 \pm \left(1.699 \cdot \sqrt{994.60/30} \right)$$

$$\text{CfI} = [535.28 \text{ kbit/s}, 556.85 \text{ kbit/s}] \quad \text{where } 100(1 - \alpha)\%\text{CfI}$$

The value, $t_{\alpha/2, n-1}$ is taken from Student t-distribution which is used when sample size is relatively small and distribution of the mean is unknown. "$n-1$" is the degree of freedom and $1-\alpha$ is the confidence level [78].

The average throughput and the 95% confidence interval of the simulation results are 547.07 kbps and (535.28, 556.85) respectively.

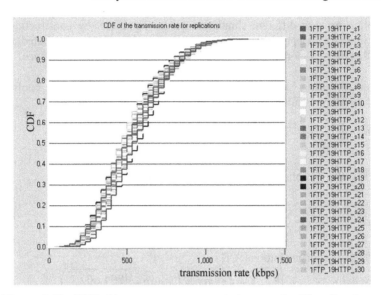

Figure 5-11: CDF of instantaneous transmission rate for 30 replications.

Further, the probability distribution for each replication is calculated and shown in Figure 5-11 which provides a broader overview of the instantaneous transmission rates.

5.1.5.4   *Comparison of Simulation and Analytical Results*

The analytical and simulation results are compared in this section. First, an analysis of the CI arrival distributions obtained by the simulation results is given in order to validate the assumption that was taken for the analytical approach. A trace from a simulation was taken for the CI arrivals and compared with the geometric distribution in logarithmic scale which is shown in Figure 5-12. The results show the close agreement of selecting a geometrically distributed arrival process for the analytical results.

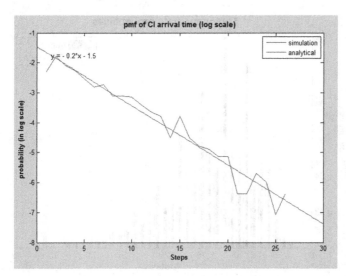

Figure 5-12: Comparison of CI arrivals pdf with the geometric distribution

Secondly, the stationary state probabilities are compared between the simulation and the analytical results as shown in Figure 5-13. One peak of the state probability (state 36) which does not occur in the analytical results appears in the simulation. It is mainly due to the fact that the UE tends to transmit with a very high transmission rate at the beginning of the simulation and receives a congestion indication due to the arrival of other users in to the system. Therefore, at the transient phase of the simulation, the UE often tends to stay in higher states compared to the UE movement is in the steady state phase. However, Figure 5-13 confirms that both the simulation results and the analytical results have a good match for all other state probabilities.

At last, the average throughput of the analytical model is compared with the simulation results with the 95% confidence interval. The average throughput calculated by the analytical model lies within the given confidence interval and is shown in Figure 5-14.

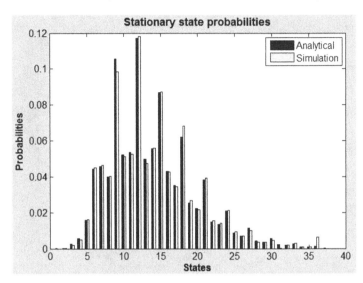

Figure 5-13: The stationary state probabilities results

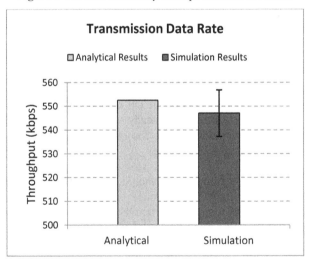

Figure 5-14: Comparison between the analytical and the simulation result

The above comparison confirms a close agreement between simulation and analytical results.

5.1.5.5   *Effect of the Reduction Factor β*

All the above analyzes were performed using default parameters for the CC algorithm which were listed in the simulation configuration of the previous section. When considering the performance, the parameters for the CC algorithm have to be well tuned. In this section, the effect of the rate reduction factor which is the key parameter for the CC algorithm is investigated using the simulation and the analytical model. The analysis is carried out by changing the β value from 0.15 to 0.25 in steps of 0.05.

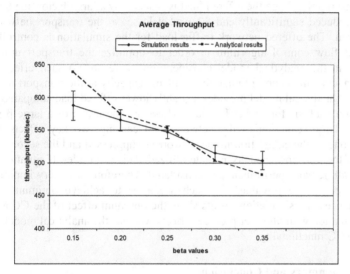

Figure 5-15: Average throughputs with 95% confidence interval

The average simulation throughputs with the 95% confidence interval for each β value were compared with the analytical results. Figure 5-15 shows the average throughput of the simulation results and the analytical results for different β.

At the beginning of the graph, for lower β values, the simulation and analytical results deviate from each other significantly. The average throughput of the simulation is lower compared to the analytical model. The reason is that from the simulation point of view, the congestion control data reduction is not sufficient for small β values in order to control the congestion of the transport network promptly. Therefore, a large number of packet losses occur at the transport network and the higher layers such as RLC and TCP start reacting to the transport congestion. Due to the additional retransmissions triggered by the RLC layer, the offered load is further increased in a short period of time which results in worsening congestion at the transport network. Lastly TCP congestion control starts controlling the data flow by limiting the offered load to the network. As a result of these subsequent effects in the higher layers, the achievable effective user throughput is significantly reduced for the simulation. However, in the analytical model such higher layer protocol effects are not considered. Hence, the average throughput differs from the simulation results.

For the range of β values from 0.20 to 0.30, both the analytical and simulation results are in close agreement. When the β value increases (such as 0.30 or 0.35), the effect of the congestion control is very aggressive and the offered load is controlled tightly. Since the offered load is reduced by a large factor, the network load is reduced significantly and the packet losses at the transport network are minimized. The offered network traffic load for the simulation is controlled by the TCP flow control algorithm in order to optimize the transport utilization whereas in the analytical model consideration, there are no such effects from higher layers which can optimize the available capacity at the transport network. Hence the analytical model provides a slightly lower offered load compared to the simulation analysis for higher β values. However, the usage of a large β values causes a large reduction of the network load causing two key effects in any system: one is, the congestion in the network is suppressed and the second is, the usage of the transport capacity is underutilized. This is a trade-off situation when the overall network performance is considered. Therefore, it is very important to choose appropriate β values for such scenarios to achieve optimum overall performance. The simulation results show the optimum effect of the CC activity in combination with all other protocol effects whereas the analytical model shows only the CC functionality.

5.1.6   Summary and Conclusion

An analytical model which is based on the Markov property has been developed in order to analyze effects of the comprehensive TNL congestion control algorithm. It has been tested and validated by using the HSDPA TNL simulator. The results closely match with an acceptable error margin. The error is mainly due to the fact that the simulation model consists of all higher layer protocol effects in addition to the congestion control functionality.

In general, the user MAC-d flow is not always in the congestion control state throughout the simulation time. The MAC-d flow alternates between the flow control phase and the congestion control phases depending on the availability of the network capacity and on the bursty nature of the user traffic. Therefore, the flow control state functionality should also be included in the analytical model in order to model the general behavior of MAC-d flow in the HSDPA network. The next section presents a comprehensive analytical model that includes both the flow control and the congestion control functionalities.

5.2   Analytical Modeling of FC and CC

The complete functionality of the congestion control has been analytically modeled, analyzed and validated using simulation results. In this chapter, both the flow control and the congestion control functionalities are modeled for each user flow. In the most general case, a user flow over a limited Iub interface alternates between flow control and congestion control states depending on the varying network capacity which is shown in Figure 5-16.

According to the explanation given in chapter 4, the flow control is an independent algorithm which estimates the average data rate over the air interface periodically for each user in the cell. The algorithm considers the channel state of the UE over a period of time which is called cycle time. The default value of the FC cycle time for the given HSDPA system is 100 ms which exhibits the optimum performance during simulation analysis and field tests [12]. Figure 5-16 shows an example scenario of the user flow behavior in FC and CC phases. It depicts the estimated FC credits for each cycle time.

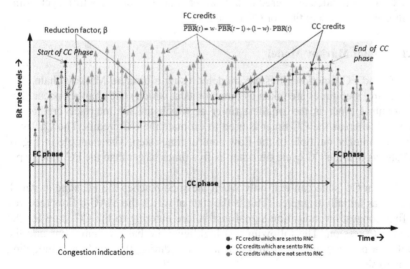

Figure 5-16: FC and CC functionalities

The HSDPA credits which are sent from RNC to eNB alternate between FC credits and CC credits. FC credits are the data rate estimated by the FC algorithm whereas CC credits are calculated by the CC algorithm. When there is no congestion in the system, all MAC flows operate in the flow control state and the FC algorithm grants credits over the transport network. The transport network congestion occurs due to the bursty nature of the HSDPA traffic. The congestion control algorithm detects and controls the input traffic flows to the transport network in order to minimize adverse effects of the congestion. Each user flow operates in congestion control state until the congestion is eliminated. During this period, user input data flow is regulated according to the CC credits which are determined by the CC algorithm. More details about these algorithms flow control, congestion detection and congestion control have been discussed in chapter 4.

The flow control and congestion detection modules issue independent signals to the congestion control module and the CC module decides the allowed credits for the congested MAC-d flow. The CC module immediately sends a Capacity Allocation (CA) message which includes the credits and interval information to the RNC in order to control the data transmission to the transport network. This CA message sending is performed periodically. The data rate of a particular user flow is constant within this period and hence the burstiness over the transport network is also significantly reduced. When a flow is in the CC state, the data rate increases by a lower rate than in the FC state. Therefore, during the periods which the CC module calculates credits are larger than the flow control cycle time which is called CC step time or CC cycle time.

5.2.1   Joint Markov Model

The CC and FC functionalities are modeled using a discrete Markov chain. The flow control step times are taken as time points at which state changes occur. The flow control timer is the smallest timer at which the air interface credits are estimated whereas the CC timer is used to estimate the CC credits. The CC timer step is an integer multiple of the flow control timer step. The example in Figure 5-16 shows that the CC cycle time step is five times larger than the FC time step.

The flow in CC state can exit only when the state reaches the FC state at which the CC was initially triggered. This is the state at which the MAC-d flow received the first congestion indication signal just before entering to the CC state for the current CC session. The following list highlights the basic assumptions and the simplifications that are taken into consideration during this analytical modeling and analysis.

- The interarrival times of CI signals are independent and identically distributed.

- The number of users in the system is constant.
- Constant transmission delay is assumed for capacity allocation (CA) messages which indicate the credits to the RNC (for HSDPA) to trigger the offered load to the transport network.
- Per-user buffer management at eNB is not considered for the FC algorithm.

When there are many users in the HSDPA system, the offered load varies significantly. Furthermore, the user air interface channel quality also changes with the ongoing time. There are many other factors such as TCP and RLC protocol functionalities which influence the load at the transport network level. The congestion can occur at any time in the limited Iub network due to the bursty nature of the offered traffic. Therefore, the first assumption is that the interarrival time of CI detection signals can be considered as independent. The second assumption is mainly due to the design of the MAC scheduler. The capacity allocation messages are sent through the uplink control channel and it is assumed that UL signaling has sufficient capacity and also a higher priority compared to the other flows of the UL traffic. Therefore, the delay for this uplink transmission can be assumed as constant which is listed as the third assumption. During this analytical modeling, it is assumed that sufficient data is in the Node-B buffer to satisfy the demanded data rate of each user flow at the air interface. Therefore, the Node-B buffer management is not considered for this analysis. The FC and CC algorithms work independently and the flow control considers the air interface fluctuation by determining the FC credits whereas the congestion control monitors the load over the transport network. Therefore, the FC and CC functionalities are modeled independently during this analytical approach.

5.2.1.1   *States Representation*

The state of the Markov model are represented by three non-negative integers i, j, and k which are defined below. The notation [i, j, k] is used to represent individual states throughout this chapter.

1. The bit rate level in the FC state denoted by i: when a MAC-d flow is completely in flow control state, the FC can vary from the lowest to the highest bit rate level based on the credit estimation of the radio interface. There are "m" such levels, thus i = 1, 2, 3,...,m. This remains constant throughout a CC session at the last FC bit rate level corresponding to the preceding FC session.
2. The bit rate levels in the CC state denoted by j: the CC state bit rate levels can theoretically achieve all FC state bit rate levels from the lowest to the highest. Thus, j is taken as 0 if the state is within a FC session and is equal to the CC bit rate level for states within a CC session Thus, j = 0, 1, 2, 3,...., m.

3. The time step count denoted by k: the CC timer is selected as an integral multiple of the FC timer. The largest value of k (denoted as n) is defined by dividing the CC timer by the FC timer. In the state representation, k spans from 1 to n with integer values. k=0 is a specific state that represents the state within the FC session whereas all other integer k represent states within the CC session. For the given analysis, n is set to 5.

5.2.1.2 *Markov State Model*

The state representation of the Markov model for the HSDPA CC and FC functionalities is shown in the Figure 5-17 along with the state transitions. There are four categories of state transitions.

- FC to FC transition:
  FC to FC transitions occur when there is no congestion in the system. States represent the UE credits which are allocated based on the varying air interface capacity. Only the state variable i changes in such a transition.

- FC to CC transitions:
  When congestion starts, a CI signal arrives and the state changes from FC state to CC state. This action always causes a rate reduction by jumping into a lower state from a higher state in the model. In a transition of this nature, the state variable i remains unchanged, j changes from 0 to a value less than or equal to i and k changes from 0 to 1.

- CC to CC transitions:
  During the MAC-d flow in the CC, the current state can only change based on another CI arrival or at the CC timer expiry. In the first case, state changes occur from the higher state to the lower state (j changes to a lower value) and in the second case, state changes from the current state to the next higher state (if k is less than 5, j remains unchanged. k increases by 1 untill it equals to 5 otherwise j increases by 1 and k is again reset to 1).

- CC to FC transitions:
  This is the transition that changes the state from CC to FC (occurs after reaching the CC state of [i,i,5]). This transition occurs when the system overcomes congestion in the Iub network. Here i changes to a value within the range 1,...,m and both j and k are reset to zero.

When the MAC-d flow is in the FC state, all state transitions occur between state [1,0,0] to [m,0,0] along the horizontal axis and the system does not experience any congestion in the transport network. Any FC state in the model can be reached from any other FC state in the horizontal axis. Self-transitions are possible at each FC state; they occur due to equal credits estimation in consecutive cycle times.

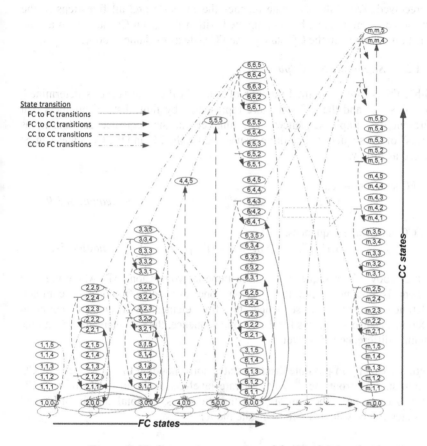

Figure 5-17: State representation of the Markov mode

The vertical transitions represent the CC state transitions. The second number of the state representation gives the notation for the CC bit rate level. For each bit rate level there are k step transitions. When no congestion indication signals are received, the CC state transitions occur due to the timer expiry and the CC bit rate levels are incremented until they reach the bit rate level corresponding to the last FC state immediately prior to the transition to CC session. After reaching this level, the MAC-d flow immediately exits from the CC state to a FC state.

The third number of the state representation denotes the step transitions for a single CC bit rate. During these step transitions, there are no changes of the bit rate levels but only the step advancements to the next step if no further CI arrivals are received. When the CC state reaches the FC state and all five steps in the considered example have been completed within the given CC state then a state transition occurs from the CC state to the FC state as explained above.

5.2.1.3   *Size of the State Space*

Within this section, the total number of states in the state space is determined. The FC states and the CC states are represented by finite lists of numbers that represent the simple sequences. The numbers in the sequence are known as "terms" of the sequence. The finite sequences for the FC and the CC states can be derived as follows.

The FC state finite sequences:
$$1,2,3, \ldots \ldots, m \overset{\text{def}}{=} (n)_{n=1}^{m} \qquad \qquad equation\ 5\text{-}9$$

The CC state finite sequences:
$$5,10,15, \ldots \ldots, 5m \overset{\text{def}}{=} (5n)_{n=1}^{m} \qquad \qquad equation\ 5\text{-}10$$

Each of above finite sequences is an arithmetic progression (AP) where the next element is obtained from the previous one by adding a constant common difference. For the first sequence, the first element is 1 and the common difference is also 1 whereas for the second sequence, the first element is 5 and the common difference is 5 as well.

There are exactly m possible discrete states corresponding to FC sessions. A CC session starting from the ith FC bit rate level can have up to 5i different CC states. According to the sum of the arithmetic series, the total number of FC and CC states can be calculated for the state model.

The total number of states,
$$m_t = m + \frac{[m \cdot (m+1) \cdot 5]}{2} \qquad \qquad equation\ 5\text{-}11$$
$$= \frac{m}{2} \cdot (7 + 5m)$$

For example, if m is 48 (eff_BW/step = $2.0*10^6*0.8/33.6*10^3$, where BW = 2Mbit/s, effective BW is 80% of total BW without protocol overhead and step height is 33.6 kbit/s) then the total number of states is 5928. This means for a single MAC-d flow, there are 48 bit rates which can be calculated according to the air interface capacity of the user channel which leads to the total number of 5928 states. The transition probability matrix is a square matrix of this size.

5.2.1.4 Input Parameters of the Analytical Model

The analytical model uses two input parameters for the evaluation. One is the FC state stationary probability distribution and the other is the CI arrival probability distribution.

As explained in chapter 4, the exact air interface modeling is not the main focus of this work. Therefore, a trace of UE air capacity in terms of the number of MAC-d packets per TTI was taken from a dedicated radio simulation. From the traces, the stationary state probabilities are derived for the flow control states for both simulation and analytical models.

The stationary probability matrix can be written as

$$PBRm = [pbr_j]_{1 \times m} \quad \text{where } j = 1,2 \dots m \qquad \textit{equation 5-12}$$

where m is the total number of bit rate levels for the FC and $PBRm$ represents the flow control bit rate level stationary probability matrix. pbr_j is the jth state probability.

The second input parameter is the CI arrival probability. This represents the probability of CI arrivals for a given interval (FC cycle time). The CC cycle time is always greater than the FC cycle time. The maximum number of CI arrivals for a given FC interval is restricted by the safe timer which is used to avoid frequent signaling over the uplink channel during congestion in the transport network. It is assumed that d_{max} is the maximum number of CI arrivals within a given FC interval. The CI arrival probability matrix for a given FC interval can be represented by A_{ci}:

$$A_{ci} = [q_i]_{1 \times (1+d_{max})} \qquad \textit{equation 5-13}$$
$$\text{Where } q_i = \text{Pr}(\text{exactly } i \text{ CI signals during } \Delta T),$$
$$i = 0,1, \dots d_{max}$$

5.2.1.5 Transition Probability Calculation

The transition probabilities of the Markov model are derived using above input parameters. There are four transition domains which are analyzed during the state modeling. Each of these transition probabilities is calculated in this section.

- The state transition probabilities for state changes within a FC session can be represented by P_{fctofc}.

$$P_{fctofc} = [p_{ij}]_{m \times m}$$
$$p_{ij} = \Pr[FC(t + \Delta T) = j \mid FC(T) = i]$$

equation 5-14

According to the Markov property, all the past history is represented by the previous state, i. Further, the relationship between past and current states depends on the equation 4-2.

A FC session continues as long as no CI signal arrives. This is taken into account by multiplying p_{ij} by the probability for zero CI arrivals within the time step. The state transition probabilities within the FC session are represented by $P_{(i,0,0)(j,0,0)}$

$$P_{(i,0,0)(j,0,0)} = q_0 p_{ij}$$

where, q_0 is the probability of no CI arrivals

occurs within a given FC interval

equation 5-15

- The transitions from a state within a FC session to a state within a CC session occur due to the arrival of a congestion indication. The number of CI arrivals can be more than one. The maximum number of CI arrivals is d_{max} and the state transition probabilities, $P_{(i,0,0)(i,l,0)}$ can be derived.

$$P_{(i,0,0)(i,l,1)} = P_{(i,0,0)(i,l,1)} + q_n$$

$$\forall l = i \times \alpha^n; \quad n = 1 \dots d_{max}$$

equation 5-16

where, $\alpha = 1 - \beta$, β is the reduction factor, q_n is the probability of n CI arrivals within a given FC interval.

- Two types of transitions can occur within a CC session. If there are no CI arrivals, state transitions can occur due to timer expiry (step timer has the same interval as the FC timer, however a different term is used in CC states). The transition probabilities are represented by the transition probability matrix $P_{(i,j,k)(i,j,k+1)}$. Here, the steps which are represented by k can change from 1 to 5. Up to step 4, there is no change in bit rate levels. At step 5, state changes occur to the next higher level of the CC bit rate level.

$$P_{(i,j,k)(i,j,k+1)} = q_0 \quad \text{for} \quad k = 1,2,3,4$$

$$P_{(i,j,k)(i,j+1,1)} = q_0 \quad \text{for} \quad k = 5$$

equation 5-17

- The second type of transition is caused by an arrival of a CI signal. The state can change from the current CC state to a CC state with lower bit rate level due to the arrival of the CI signals. When multiple CI arrivals occur in a single step time period, the current state can fall back to a CC state which has much lower bit rates. It is now considered that it fell back from the CC state j to the CC state l due to the n CI arrivals.

$$P_{(i,j,k)(i,l,1)} = P_{(i,j,k)(i,l,1)} + q_n$$

equation 5-18

$$\forall l = j \times \alpha^n; \quad n = 1 \ldots d_{max} \text{ and } k = 1 \ldots 5$$

where, $\alpha = 1-\beta$, β is the reduction factor, q_n is the probability of n CI arrivals within a given FC interval.

- The state transitions from the CC state to the FC state occur due to the absence of CI arrivals for a long period of time. When the CC bit rate level reaches the FC bit rate level at which the MAC-d flow started to move from FC session to the CC session, the current CC state changes to a FC state. It is assumed that i is the FC bit rate level at which the current CC session started and l is the next FC bit rate level which is entered to. The new FC bit rate level can be any level that is selected based on the current air interface channel capacity. All these transition probabilities can be represented by the transition probability matrix $P_{(i,i,5)(l,0,0)}$.

$$P_{(i,i,5)(l,0,0)} = q_0 \, P_{il}$$

equation 5-19

l is the next FC state and i is the starting FC state

before the MAC-d flow enters the CC state

After completing the calculation of all state transition probabilities, the transition probability matrix is given below.

$$P = \left[p_{ijk} \right]_{n_t \times n_t}$$
where $i = 1,2,3, \ldots , n_{fc}$

$$j = 1,2,3, \ldots , n_{cc}$$
$$k = 1,2,3, \ldots , n_{st}$$

equation 5-20

The stationary state probabilities are calculated by solving equation 5-21 along with the condition that sum of all steady state probabilities equals one. In order to solve this linear system equation, the same approach is used that has been explained in section 5.1.3.

$$\pi = \pi \cdot P \qquad\qquad \text{\textit{equation 5-21}}$$

where π denotes the state vector, $[\pi_0, \pi_1, \pi_2, \pi_3, \dots, \pi_{nt}]$.

The throughput is calculated using the following equation 5-22. The bit rate steps are defined in terms of 33.6 kbps.

$$Average\ throughput\ = 33.6 \cdot \sum_{i=1}^{n_t} i \cdot \pi_i \ \text{kbit/sec} \qquad \text{\textit{equation 5-22}}$$

5.2.2   Analytical and Simulation Parameter Configuration

The simulation and the analytical models use the same set of parameters.
- FC cycle time = 100 ms
- CC AIMD cycle time = 500 ms
- FC and CC step size = 33.6 kbps (1packet/10ms)
- $\beta = 0.25$
- Safe timer = 80 ms
- ATM bandwidth = 2Mbit/sec

In addition to the above, two input parameters which will be discussed next are taken from two separate simulation models. They are the probability distribution of the radio channel capacities in terms of MAC-d/TTI and the mean CI arrival rate per step. The radio traces are taken from the dedicated radio simulation model whereas the mean CI arrivals are taken from a fast queuing simulator which will be discussed in chapter 5.2.5.

For the system simulation, a comprehensive OPNET-based HSPA simulator (chapter-3) which has been designed and developed by the author is used. The simulation specific parameters are listed as follows.

- Traffic models: FTP and HTTP traffic models [23].
- User constellation: 18 users, 1 FTP user who downloads a large file during the simulation time and uses the probability distribution used for the analytical model. All other 17 users generate HTTP traffic.
- The simulation duration is 2000 sec and 32 replications are used to determine the confidence intervals.

5.2.3   Input Distribution of Radio User Throughput

The input probability distribution of the radio throughput is derived from the common trace file which includes the air interface channel capacity in terms of number of MAC-ds/TTI. This input distribution is used both for the simulation and the analytical model. The probability mass function of the input is given in Figure 5-18.

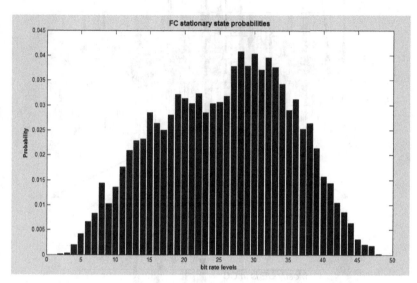

Figure 5-18: Probability mass function of the number of MAC-ds

The average air capacity of the user considered in the figure is about 907 kbps. This is the case without considering the HSPA access network. When such a user operates in a real system with limited Iub capacity shared with other users, the achievable data rate of the users is further reduced due to transport congestion. The next section presents the simulation results in which FC and CC algorithms are included within the detailed system simulator in order to achieve the aforementioned achievable user throughput over the transport network.

5.2.4   Simulation Results Analysis

The HSPA OPNET simulator is configured with the above common set of parameters as well as simulation specific parameters. 32 simulations were run using different random seeds. The instantaneous transmission rate (in green) and the average transmission rate (in blue) over the model time of a single simulation run are shown in Figure 5-19 as an example scenario.

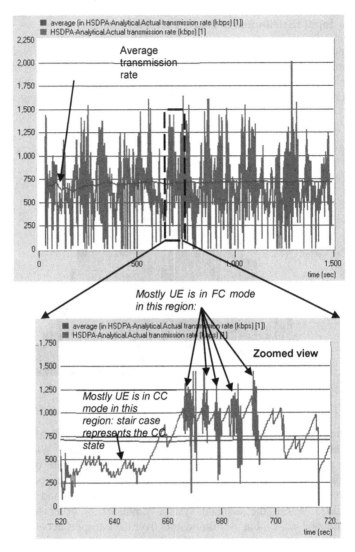

Figure 5-19: Instantaneous and average transmission rates

It is shown that at 1500 seconds of model time, the average throughput variation becomes stable and the system approximately achieves steady state behavior. Due to the bursty nature of the HSDPA traffic over the transport network (Iub), the output data rate fluctuates inside a large range. Furthermore, Figure 5-19 also presents a zoomed version of the relevant portion from the figure giving a closer look. The user flow fluctuates between two domains, the flow control domain and the CC domain. At the beginning of the zoomed window in Figure 5-19, the MAC-d flow follows the saw-tooth-like regular pattern which is typical for a CC session. The data rates over the Iub within CC sessions are lower compared to those in FC sessions and that reduces the overload situation in the transport network. At the middle of the zoomed window, the figure shows high fluctuations of the output data rates. This happens when the user flow completely operates in a FC session. If there is no congestion in the transport network, the flow completely follows the air interface channel capacity which varies over a large range.

Mostly congestion occurs when many users who have a very good radio channel capacity try to access the limited transport network with full capacity at the same time. Therefore, such situation causes many packet losses at the transport network and often many RLC and TCP retransmission triggers. Before this case occurs, the CC module detects the upcoming congestion and takes action to limit the data rates of some users by changing the state to CC state from FC state.

The probability distribution of the instantaneous transmission rate is shown in Figure 5-20.

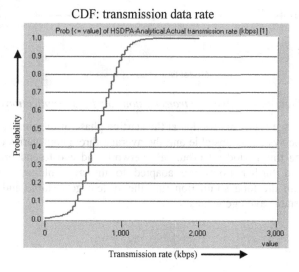

Figure 5-20: CDF of the instantaneous data rate

In order to analyze the statistical quantities such as the confidence interval for the simulation results, 32 replications of the simulations were done. The average throughput and the probability distribution (cumulative probability distributions) of throughputs for those simulation runs are shown in Figure 5-21 and Figure 5-22 respectively.

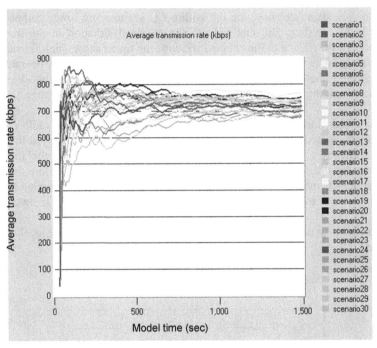

Figure 5-21: Average transmission rate for 32 replications

The time average figure shows that at the starting phase of the simulation (warm-up phase), the system is unstable and the average throughput varies over a large range. During this period all protocols (network and end-to-end protocols) are initialized and further flows are adapted to the available system capacity. However, with the long simulation run, the system gets stable and the results converge to stable average values.

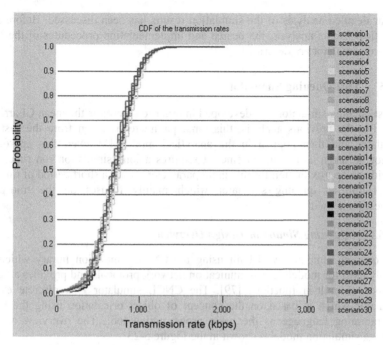

Figure 5-22: CDF of the instantaneous transmission rate for 32 replications

The confidence interval is calculated by taking the means of the above replications. For this calculation, the warm-up period of the individual replications is not included. The instantaneous output data rates (simulation output) from 500 sec to 1500 sec were chosen to calculate the mean throughput of each replication which is shown in Table 5-2.

Table 5-2: Mean throughputs of 32 replications

Replications	1	2	3	4	5	6	7	8	9	10
Mean	713	697	700	688	743	676	740	726	720	689
Replications	11	12	13	14	15	16	17	18	19	20
Mean	723	706	734	726	743	763	704	715	724	749
Replications	21	22	23	24	25	26	27	28	29	30
Mean	702	742	730	717	681	732	718	729	736	710

The mean and standard deviations of the transmission data rate are 719.60 kbps and 21.13 kbps respectively. The 95% confidence interval is [713.05 kbps, 726.16 kbps].

So far detailed analysis of the simulation results has been discussed. Before the analytical results analysis, the design and implementation procedures of the fast queuing simulator are described.

5.2.5   Fast Queuing Simulator

A fast queuing simulator was developed in order to determine the mean CI arrival rate. For the previous analysis, this input parameter is taken from the system simulator and then applied in the analytical model. However, this approach cannot be taken as a solution since it requires a long simulation run time. To provide an approximation for this input parameter within a short period of time, a fast simplified queuing simulator which mainly focuses on the transport congestion is used.

5.2.5.1   *Queuing Simulator Design Overview*

The queuing simulator was built using the CNCL simulation library which is widely used for modeling communication network protocols and provides a large number of built-in functions [79]. The CNCL simulator is a discrete event simulator which is based on the concept of object orientation using the C++ programming language as the basic programming tool. The overview of the queuing simulation model is shown in the Figure 5-23.

Figure 5-23 shows two main parts of the simulation model: the TNL source manager network and the transport network. The TNL source manager generates traffic to the transport network. The user traffic flows are independent and are provided by a large number of TCP sources depending on the configured application profile. For example, FTP uses a separate TCP connection for each file download whereas web traffic uses one TCP connection for the download of each page.   Each data flow which consists of one or many TCP sources or connections uses one RLC buffering unit with a sufficiently large capacity in the RNC entity. Since there are "n" user flows, Figure 5-23 shows "n" RLC buffers in the RNC. In many cases RLC buffer occupancy rapidly increases when congestion occurs in the transport network. The maximum RLC buffer capacity is directly controlled by the TCP end-to-end flow control scheme and also depends on the number of simultaneous connections. All application data is stored at the RLC buffer until they are transmitted to the transport network.

Data is sent from the RLC buffer to the transport network according to the trigger from the CC module. As described in chapter 4, the CC module takes two input triggers: flow control and congestion indication, and decides the output data rate by considering the transport utilization. All these input triggers are independent within each data flow.

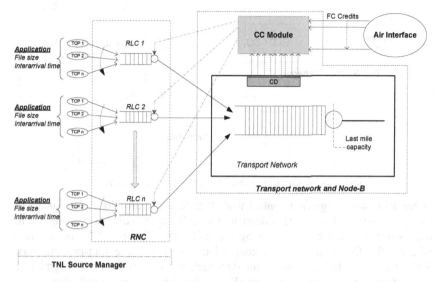

Figure 5-23: Queuing simulation model

The transport network is modeled with a queue and a server. The serving rate is determined by the last mile link capacity which is the bottleneck of the transport network. The buffer unit which has a limited capacity represents all buffers in the transport network. In general, the transport network is dimensioned to keep the packet loss ratio under a certain maximum limit which is for example 1% of the total packet transmission. The congestion control algorithm minimizes the transport packet losses while optimizing the transport utilization. The complete congestion control procedures are described in chapter 4.

A large number of frame protocol data packets from independent application sources arrive at the transport buffer. The decision regarding the next transmission opportunity of these packets is taken using independent transport triggers.

5.2.5.2   *Queuing Simulator Implementation*

Figure 5-24 shows the block diagram of the queuing model implementation overview in the queuing simulator. As shown in Figure 5-24 the implementation is performed by dividing the complete model into two parts: the TNL source manager module which generates and controls the traffic sources, and the transport network.

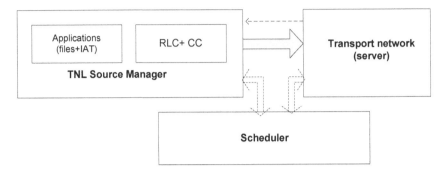

Figure 5-24: Queuing simulator implementation overview

When RLC and congestion control work together, all transport effects such as a small number of losses are hidden to the TCP layer. All lost packets at the transport network are recovered by the RLC layer before TCP notices them whereas the CC minimizes the congestion situation at the transport network. Further the RLC buffers receive the data packets which come from several TCP connections. The packets are stored in the RLC queue until they get a transmission opportunity. Therefore the TCP slow start effects are mirrored by the RLC controller unit to the transport network. The modeling of the complete TCP protocol functionality creates again a complexity to the fast queuing simulator and it may require a longer simulation time for the simulations as the detailed system simulator does. On the other hand, when the aforementioned details are considered, the modeling of TCP protocol does not have a significant impact to the output of the fast queuing simulator. Therefore, the modeling of the complex TCP protocol is excluded within the source modeling of the fast queuing simulator. However during application modeling all other parameters such as file size and interarrival distributions are correctly modeled in the fast queuing simulator which is similar to the traffic modeling in the detailed system simulator.

The flow chart of the TNL source manager is shown in Figure 5-25 and Figure 5-26. All user traffic flows start with the initialization process and start generating application packets (APP_JOB) with a random offset. The user flow can be either in ACTIVE state or INACTIVE state. When UEs have data to be sent to an application input queue, then the user is in ACTIVE state otherwise they are in INACTIVE state.

The size of the FP JOB (or FP PDU) is determined according to the credits which have been granted by the CC controller in the RNC.

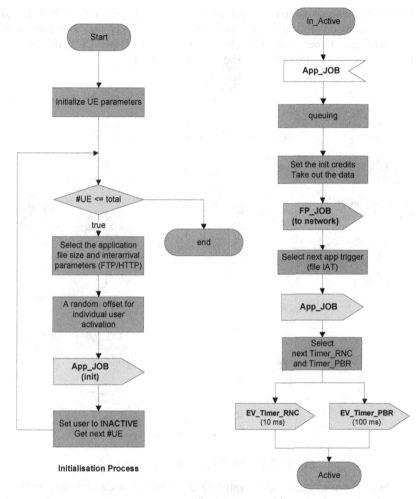

Figure 5-25: TNL source manager (part 1)

The FP JOBs are transmitted at the expiry of every RNC time interval which is set by the EV_Timer_RNC event trigger. For example the default value of the timer event is set to 10 ms. The credits are updated every FC cycle time if the user flow is not in congestion. This cycle timer expiry is also triggered by EV_Timer_PBR in the TNL source manager module, the default value is set to100 ms. When a user flow is causing congestion in the transport network, the CC instead of the FC starts to control the sending data flow. In this case, the credits are calculated by the CC algorithm with longer cycle time in order to reduce input traffic to the transport network. The default CC time is set to 500 ms for this analysis which is triggered by EV_Timer_CC event. When the UE flow is in ACTIVE state, there are three timer events and one Job event: App_JOB, EV_Timer_RNC, EV_Timer_PBR and EV_Timer_CC which occur in the TNL source manager block as shown in Figure 5-26.

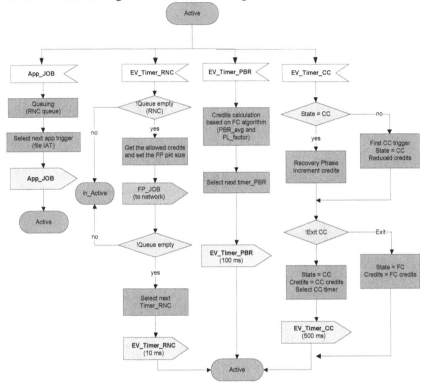

Figure 5-26: TNL source manager (part 2)

All above event triggers are activated on a per flow basis and continuously run when the flow is in ACTIVE state of the TNL source manager. Depending on the timer events the respective algorithms start processing the input data. For example if the EV_Timer_PBR timer arrives, the FC algorithm starts the calculation procedures in order to update the credits. Similarly when EV_Timer_CC timer is received, credits are calculated according to the CC algorithm.

All FC and CC algorithm functionalities are depicted as blocks in the flow chart in Figure 5-26. The FP JOBs (or FP PDUs) are sent to the transport network continuously until the application input buffer is empty throughout the simulation period. Figure 5-27 shows the transport network (or server) functionality. It has two states: WAITING and SERVING. Based on the arrival of FP job events, the transport network starts sending data by changing the flow state from waiting to serving.

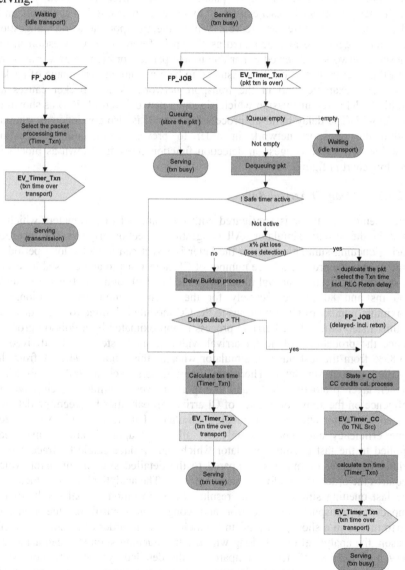

Figure 5-27: Transport network (server) functionality

Any arrival of a packet when the server is busy (or during transmission) is stored in the buffer. When the transmission duration is over based on the arrival of the EV_Timer_Txn event, the next packet in the buffer is served and the EV_Timer_Txn timer is rescheduled. In this manner transmission continues until all packets in the buffer are served. Within the transport network, congestion detection triggers are also activated as shown in figure 5-27. Congestion in the transport network is detected either due to lost packets or high delay variation of the FP packet transmission. It is assumed to be a maximum percentage of packet losses (for example 1%) in the transport network. A lost packet causes an additional RLC retransmission which triggers a new delayed FP JOB as shown in figure 5-27. The delay build-up process [chapter 4] is also continuously running within the transport network in order to prevent forthcoming congestion situations. All these congestion detection functionalities are shown as blocks in the flow chart in figure 5-27.

5.2.5.3   *Mean CI Arrival Rate*

The queuing simulator is configured with the same set of parameters which is used in the system simulator. All congestion detection triggers are recorded during queuing simulation. Since the latter is fast it can model a long period of time in order to provide a large number of CI indications to achieve stable results of computing mean CI arrival rates. Since the FC cycle and safe timers are set to 100 ms and 80 ms respectively for the given example configuration, the maximum number of CI indications to the CC module is limited to one indication within a step time. The CI arrival process is approximated by a Poisson process, hence the probability of no CI arrival within a single step can be derived as 95.83% from the fast queuing simulator whereas this value is 94.74% from the detailed system simulator. There are many factors such as traffic modeling, higher layer protocol effects and transport network settings which have an influence on the above difference of CI arrival probabilities between the detailed system simulator and the fast queuing simulator. Due to the trade-off between time efficiency and model complexity, all above approximations have been applied to the fast queuing simulator which can produce stable CI trace results within the order of minutes compared to the detailed system simulator which requires the order of days for the same outcome. The analytical model which uses the fast queuing simulation trace results is able to produce final results of the impact of transport flow control and congestion control on the transport performance in a short period of time which is also in order of minutes. For this reason, the analytical model along with the fast queuing simulator can achieve a very high overall efficiency compared to the detailed system simulator. More details about the analytical model results and the fast queuing simulation aspects are also discussed in the results comparison chapter 5.2.7 by highlighting the key differences.

According to the equation 5-13, the probability of CI arrivals within a single step (q_I) and the maximum number of CI arrivals within single step (d_{max}) parameters which are given by the fast queuing simulator are 0.0417 and 1 whereas the comparison values of 0.0526 and 1 are provided by the system simulator respectively. There are several reasons for difference of the probability of CI arrivals between the simulation and the fast queuing simulator which will be discussed in chapter 5.2.7.

5.2.6   Analytical Results Analysis

In order to evaluate the analytical model results, the MatLab tool is used. The analytical model uses fast queuing simulation results as input parameters along with the common set of parameters which have been discussed in chapter 5.2.2.

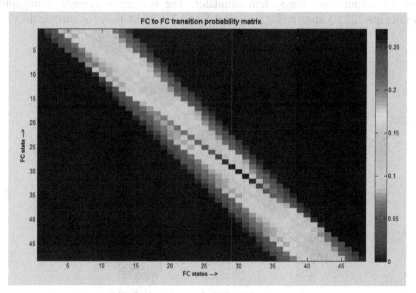

Figure 5-28: Transition probabilities from FC to FC.

For this analysis, the CC timer which is configured to 500 ms is selected 5 times larger than the FC timer (number of steps k = 1, 2, 3, 4 and 5). The FC timer of 100 ms is considered as the smallest time interval. There are 5 steps of FC triggers within one CC trigger. According to these configurations, the transition probabilities of the Markov model have been calculated. For example, the transition probabilities from FC state to FC state are shown in Figure 5-28.

The state changes from lower to higher states have a very rare probability due to the averaging equation 4-2. During the PBR calculation, all values of the past states are considered to be part of the previous state, therefore when the average value is calculated, a large portion of the previous values is taken for the current state calculation. Furthermore, the figure shows that higher state transitions occur between states 15 to 40, in which the range of the average radio throughput lies for this particular MAC-d flow.

The complete transition probability matrix (*equation 5-20*) for the given example scenario is a square matrix with the dimension of 5928×5928. Due to its large size no attempts are made to represent it graphically.

The average throughputs for the analytical model which are calculated using the equation 5-21 are depicted in figure 5-29. The analytical-Q setup uses the mean CI arrival rate from the fast queuing simulator whereas the analytical-S setup uses the CI output from the system simulator. The respective average throughput values are 722.63 kbit/sec for analytical-S and 748.48 kbit/sec for analytical-Q.

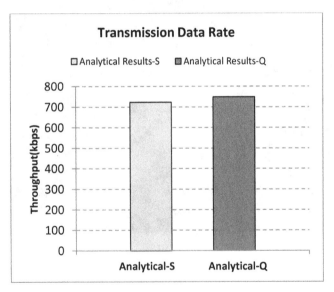

Figure 5-29. Average throughput calculated using analytical model

5.2.7   Result Comparison and Conclusion

Under this section both simulation results and analytical results are compared for two example scenarios. First, the aforementioned simulation and analytical results are compared together and the performance is discussed in detail. Secondly, for further validation of the analytical approach, another configuration by adding 10 more bursty web users to the current system is analyzed and compared with the simulation results.

5.2.7.1   *Results Comparison and Analysis 1*

Figure 5-30 shows the average throughputs which also include all transport overheads and RLC retransmissions for both the analytical and the simulation results. As discussed previously, the analytical model provides the average throughput for two different configurations analytical-S which uses the trace from the detailed system simulator and analytical-Q which uses the CI probability from the fast queuing simulator.

Figure 5-30 depicts that the analytical-S scenario provides the results within the 95% confidence interval of the simulation outcome whereas the results of the analytical-Q scenario slightly deviates from the confidence interval. This deviation is mainly due to the various assumptions made during the modeling of the fast queuing simulator such as that the CI arrival processes is assumed as a Poisson process. Despite this, the model simplifications such as the effect of the TCP sources have been neglected. Further, 1% packet loss probability was assumed at the transport in which the lost events are uniformly distributed among the packet arrivals whereas often packet losses are experienced in bursty nature in the detailed system simulator. For the traffic modeling, the fast queuing simulator assumes one object which is negative exponentially distributed for each web page by considering its average page size whereas detailed system simulator considers five separate objects which are Pareto distributed for each page.

Even though all aforementioned differences cause some influence to the final outcome, only some of them can be quantified due to the complexity of the system. For example, the fast queuing simulator assumes a fixed packet loss probability of 1% for all analysis. However, during the simulation run, the detailed simulator experiences 0.43% loss probability at the transport level for the given configuration. This clearly indicates that the fast queuing simulator increases the throughput at the transport network level due to a higher number of additional retransmissions triggered by the RLC layer to recover transport packet losses in comparison to the detailed system simulation. The average overhead due to RLC retransmissions can be quantified and is approximately 6.8 kbps in case of the fast queuing simulation for the given scenario.

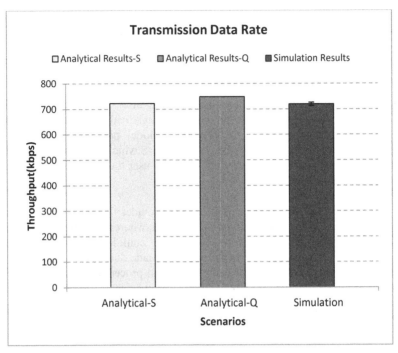

Figure 5-30: Average throughputs for both simulation and analytical results

It is shown that all above overheads cause a large difference in the throughput comparison but fewer effects on the goodput which is the throughput without transport overheads. Therefore, the goodput comparison between the detailed system simulation and the analytical-Q model shown in Figure 5-31 exhibits a closer agreement in contrast to the throughput comparison. The goodput is calculated by excluding both the general transport packet overhead which is approximately 24% and additional RLC retransmissions. In addition to the comparison of limited transport results, Figure 5-31 shows an additional result in which the ideal transport is considered. In the limited transport based scenarios, all analyzes were performed with a load factor of more than 100% to exhibit the effectiveness of the FC and CC algorithms. Therefore, the achievable goodput performance for the limited transport scenarios is greatly reduced compared to the ideal transport consideration.

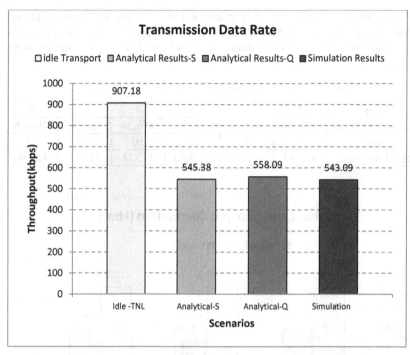

Figure 5-31: Goodput comparison among all simulation scenarios

The goodput is 907 kbit/sec when ideal transport is used and all three limited transport analyzes – analytical-S, analytical-Q and simulation – show a close agreement of the goodput performances which are 545 kbit/sec, 558 kbit/sec and 543 kbit/sec respectively as depicted in Figure 5-31. Therefore the results confirm that the analytical model along with the fast queuing simulator provides a very good outcome and is able to perform FC and CC analysis in a very short period of time which is in the order of minutes compared to the full featured detailed system simulator which requires some days for the same analysis.

5.2.7.2   *Results Comparison and Analysis 2*

As mentioned above, another example scenario is added to validate the analytical model results using detailed system simulation analysis. In order to compare with the previous analysis, this example scenario is configured by adding 10 more web users to the previous system leading to an increase of the average offered load approximately by 400 kbps. Therefore, this example scenario causes severe transport congestion which has to be minimized by further CC and FC activities.

The simulation was run with 12 replications; the average throughputs for all simulations are shown in Table 5-3. The overall average throughput and the 95% confidence interval are 457.08 kbit/sec and (442.55 kbit/sec, 471.60 kbit/sec) respectively.

Table 5-3: Average throughputs of 12 replications

Replications	1	2	3	4	5	6
Mean throughput	440.99	494.76	438.72	456.58	468.50	466.06
Replications	7	8	9	10	11	12
Mean throughput	485.43	453.03	459.10	475.18	416.65	429.96

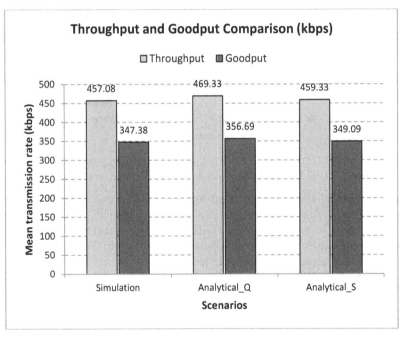

Figure 5-32: Average throughput and goodput comparison

The Analytical-Q model also uses the same parameters and number of users for the analysis. By configuration 10 additional web users in the fast queuing simulator, the simulations are performed to achieve the CI arrival probability distribution. The mean CI arrival probability within a step time is 20.0% given by the fast queuing simulator compared to the mean CI rate of 20.79% given by the detailed system simulator. The average loss ratio for the detailed system simulator is ~ 0.91% which is much closer to the value of 1% which is assumed by the fast queuing simulator. Therefore, in both cases the total number of RLC retransmissions is nearly equal and has no significant impact on the final results in contrast to the previous analysis. The results confirm that the transport level congestion is severe for this configuration compared to the previously analyzed configuration, leading to many losses and CI triggers equally in both simulators. Therefore the transport level throughputs are in close agreement similar to the goodput comparison. Both throughput and goodput results are shown in Figure 5-32.

The average throughputs for simulation, Analytical-S and analytical-Q configurations are 457 kbit/sec, 459 kbit/sec and 469 kbit/sec respectively. Both the analytical-Q and analytical-S results lie within the 95% confidence interval (442.55 kbit/sec, 471.60 kbit/sec) of the simulation results. This shows a good match between the simulation and the analytical results even at a very high congestion level in the transport network.

5.2.8   Conclusion

Within this chapter an analytical model along with the fast queuing simulator has been implemented, tested and validated. The aforementioned analysis shows that a fast queuing simulator can perform a complete FC and CC analysis within an order of minutes compared to the detailed system simulator which requires an order to days for the same analysis. Further results of the fully analytical approach (the analytical model along with the fast queuing simulator) show a good match with simulation results. Therefore the analytical model can be effectively used to analyze the effect of the FC and CC parameter Optimization providing a great flexibility along with acceptable accuracy within a short period of time. Further the model can also be efficiently used to emulate different bursty traffic environments for the performance analysis of the aforementioned algorithms. Therefore, the analytical model can be used as a cost-effective alternative solution to the time consuming detailed system simulations.

6 Conclusion and Outlook

This thesis work commenced focusing on the optimization of high speed mobile access networks to achieve best end-to-end performance. Several novel approaches have been proposed to achieve the above objectives. All novel algorithms have been implemented, tested, validated and analyzed using comprehensive simulation and analytical tools. A prominent enhancement is achieved not only for the end user performance but also for the overall network performance. The proposed solutions significantly improve the high speed mobile access network utilization by providing cost-effective benefits to the mobile network operators and the end users.

The key challenges such as developing comprehensive simulations and analytical models for high speed mobile access networks have been successfully achieved during this work. The comprehensive HSPA and LTE simulators which have been designed and developed by the author are able to test, validate and analyze all scenarios which focus on transport network performance. In addition to the thesis work, these simulators have also been used by the industrial projects managed by Nokia Siemens Network to investigate and analyze the UTRAN and the E-UTRAN performance respectively. All peer-to-peer protocols such as TCP, RLC and PDCP, and other protocol layers such as MAC-hs and the E-DCH scheduler for HSPA and the transport diffserv model and the time-frequency domain MAC scheduler for LTE have been implemented in the respective simulators according to the common framework specified by the 3GPP specification. These simulators are able to perform an extensive analysis from the application to the physical layer by providing detailed statistics. Both simulators were constructed using a modularized approach in order to enable extensions with great flexibility and enhanced scalability, adjust the required granularity for different levels of investigations, optimize for long simulation runs and select the appropriate statistics for different analyzes.

The impact of the credit-based flow control on the overall network and the end user performance was investigated and analyzed using both simulation and analytical tools. The results confirm that the algorithms achieve lower burstiness over the Iub link, lower buffer occupancy at eNB, lower FP delay and lower delay variation at the transport network in comparison to the generic ON/OFF flow control mechanism. Based on above achievements, the transport network utilization has been significantly enhanced by reducing the required capacity for bandwidth recommendations on the Iub interface of the HSPA network. Hence the end user and the overall network performance are greatly improved by optimizing the goodput (throughput at the application layer) of the network.

High speed packet traffic often has a bursty nature and leads to inefficient usage of resources for both the radio and the broadband access networks. The proposed congestion control algorithms work jointly with the credit-based flow control in an effective manner to solve the aforementioned issues. Several variants of congestion detection algorithms such as preventive, reactive and combined have been investigated and analyzed to find which of them are able to provide the best overall performance. All these algorithms have been tested and validated for the performance of the high speed mobile access network using simulation and analytical approaches. The results confirm that the combined usage of the preventive and the reactive congestion control algorithms can achieve a significantly better overall performance compared to other configurations. Such algorithms can protect the transport network from any transient behavior due to bursty traffic by increasing the reliability of the system in order to guarantee the end user and overall network performance. Furthermore, the analysis confirms that these algorithms optimize the effective transport network utilization and thus provide cost-effective benefits for both the network operators and the mobile users.

Two novel analytical models which are based on Markov chains were developed within the focus of this thesis in order to overcome issues such as long timing requirements not only for the development of the simulator but also of the simulation itself, related to the simulation approach. Out of these analytical models, one is designed to analyze the congestion control and the other is to analyze both the flow control and the congestion control functionalities. In a real system, there are some users who try to exploit the system capacity by downloading a large amount of data continuously for a long period of time. These flows are completely handled by the congestion control algorithm covered by the first analytical model in order to monitor and control their offered load to the transport network during the entire connection period. In contrast to that, the moderate users who do not greedily use the network vary their offered load in a discontinuous manner based on their time dependent requirements. Therefore such a flow is operated under both the flow control and the congestion control schemes which are investigated and analyzed by using the second analytical model. Since the analytical models require an input parameter about the congestion behaviors of the transport network, a fast simplified queuing simulation was also developed by modeling the transport network part of the access network along with the required other protocols. Due to the simplicity of the fast queuing simulator, the latter is able to provide the required input to the analytical model significantly faster than the detailed system simulator. It is shown that by applying these congestion indication inputs, the analytical model is able to perform the complete analysis of the flow control and the congestion control to evaluate the end user performance and overall network performance efficiently. The main advantage of the analytical models is that they can perform the same analysis as the detailed system simulator does within a shorter period of time, i.e. in the order of minutes compared to an order of days required by the detailed system simulator. Further results confirm that there is a good match between the analytical and the simulation approaches. Therefore, the analytical models can be effectively used to perform fast and efficient analyzes with a good accuracy compared to the detailed system simulation.

The LTE investigations were mainly focused on evaluating the end-to-end performance under different transport effects such as transport level QoS classification, transport congestion, LTE handovers and bandwidth dimensioning. By deploying suitable real time traffic models such as VoIP and best effort traffic models such as FTP and HTTP, the end user performance was evaluated. All analyzes confirm that LTE requires an enhanced traffic differentiation methodology at the transport network in order to guarantee the end user QoS effectively. For example, a very good speech quality for the end users can be achieved when the VoIP traffic is assigned a higher priority over the best effort traffic at the transport network. Further, it is observed that when HTTP traffic is mixed with large file downloads over the transport network, the web page response time is significantly increased and provides a very poor end-to-end performance. Therefore, the optimum HTTP performance is obtained when the HTTP traffic is separated and given higher priority than large file downloads at the transport network. Simulation results further emphasize that seamless mobility for end users can be provided during LTE handovers even without prioritizing the forwarded data over the transport network if RLC and TCP work appropriately recovering packet losses at the transport network.

The LTE deployment has been recently started in several countries such as Norway, Sweden and Germany. However, there are many performance issues regarding the optimum usage of the E-UTRAN network resources which are not yet being addressed. During this thesis work, the transport network investigation and analysis have been performed using few bearer types along with the service differentiation mainly at the transport network for the uplink traffic. There is no flow control mechanism considered to optimize the network performance between the MAC scheduler and the transport scheduler and to guarantee the QoS of different bearers in the packetized network. Further, such a flow mechanism should be able to provide load balancing between cells by optimizing the usage of the available transport capacity while minimizing the inter-cell interference. It is also necessary to build an adaptive congestion control technique by applying back pressured feedback – the information about varying transport capacity – from the transport scheduler to the MAC scheduler for the uplink. By deploying above improvements, the end-to-end performance and overall network performance should be investigated and analyzed in an environment where both radio and transport aspects are considered at the same time.

Appendix-A.   DSL Transport on HSPA Performance

Several investigations and analyzes have been performed in order to understand the effect of the DSL based UTRAN on the HSPA performance. Out of all these analyzes, two main investigations "Effect of DSL delay and delay variation on HSPA performance" and "Effects of the DSL fast mode operation on HSPA performance" are presented in the thesis.

The main traffic models – ETSI based web traffic and FTP traffic – that are described in chapter 4 are used for the downlink analysis whereas a moderate FTP traffic model which is described in the standard (3GPP TS 2525.319 V7.2.0) is used for the uplink analysis. The main parameters for 3GPP uplink FTP is described in Table A-1.

Table A-1: 3GPP FTP traffic model parameters

FTP traffic model parameters	
File Size	Truncated Lognormal Distribution Mean: 2 Mbytes, Max: 5 Mbytes Std. Dev.: 0.722 Mbytes
Inter Arrival Time	Exponential Distribution Mean: 180 sec

Next, all the DSL investigations and analyzes for the aforementioned two considerations are described.

A.1   Effects of DSL Default Mode on HSPA Performance

The following analyzes are carried out to characterize the effect of the DSL transport delay and delay variation on the HSPA performance. A trace file, taken from a DSL normal mode operation in which the bit errors are randomly distributed along the DSL data stream are deployed for this analysis. This configuration is also named as the DSL default mode of operation.

Three simulation configurations are selected to analyze the performance.

- **Configuration 1:** Idle DSL transport and idle DSL delay
  In this method, a high BW is allocated to the transport network which simulates a negligible delay and delay variation in the transport network.
- **Configuration 2:** Idle DSL BW and limited DSL delay
  In this configuration, the idle BW is configured for the DSL link but static DSL delay and delay variation is emulated using the given ADSL2+ trace file.
- **Configuration 3:** Uplink limited DSL BW and Limited DSL delay
  Under this configuration, the DSL uplink BW is configured to 1.23 Mbit/sec and DSL downlink BW is configured to 13 Mbit/sec.

All three configurations use 6 uplink users and 14 downlink users who are downloading large FTP files all the time. The users are configured with the FTP worst traffic model which has been discussed at the beginning of this study. In addition to the HSPA users, there are Rel'99 users in the downlink and the uplink who occupy ~2 Mbit/sec BW from each link.

According to the above configurations, no effect is expected for configuration 1 and configuration 2 for the downlink due to sufficient capacity in the downlink transport network which can fulfill the radio BW requirements. However, configuration 3 which has a limited uplink may block higher layer downlink signaling which can reduce the achievable throughput for the end users. In order to clarify above effects, important statistics for the simulation results are presented. The simulation scenarios (configurations) use the following naming: the configurations 1, 2 and 3 are respectively called "ideal", "delay_ideal" and "delay_limited".

Uplink Application Throughput

Figure A-1 shows per user application throughputs. The throughput is measured at the application layer, therefore all the effects of the lower layer are included in these throughputs. The idle scenario (configuration 1) has the best throughputs whereas the delay limited scenario has the lowest throughputs which are due to the limited uplink capacity. The fairness between the end user throughputs is excellent for the delay limited simulation scenario. The results of the configuration 2 show that there is no significant impact having small static delays over the DSL network.

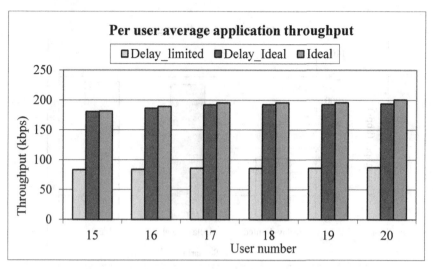

Figure A-1: Uplink per-user application throughputs

Overall application throughputs and fairness

The total application throughput and fairness are shown in Figure A-2 and Figure A-3 respectively. The overall throughput follows the same argument which is given in the per-user throughput results.

The unfairness factor is determined by using per-user application throughputs and the formula is given.

$$UF = \frac{1}{N} \cdot \sum_{i=1}^{N} \left| T_i - T_{avg} \right| \quad \text{with } T_{avg} = \frac{1}{N} \cdot \sum_{i=1}^{N} T_i$$

equation A 1

$$UF_{\_Tavg}\% = \left(UF / T_{avg} \right) \cdot 100$$

where T_i is the individual per-user throughput and T_{avg} is the average of all per user throughputs. UF is the unfairness factor.

The unfairness factors are very small values for three scenarios. The configuration 3 has the best fairness among them. However, there is no significant impact on the unfairness due to the DSL transport.

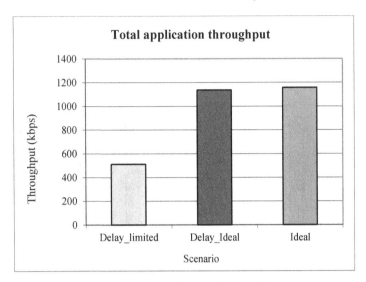

Figure A-2: Uplink overall throughputs

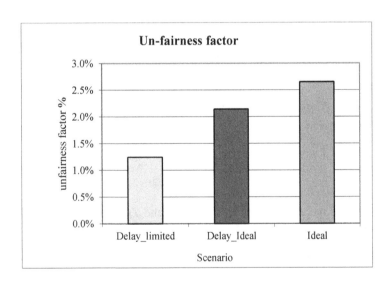

Figure A-3: Uplink unfairness factors

Downlink Overall Application Throughput and Fairness

The total downlink application throughput and fairness are shown in Figure A-4 and Figure A-5 respectively. All simulation configurations show a similar HSPA end user performance for fairness and end user throughput and hence it can be concluded that having a limited uplink does not affect to the HSPA performance.

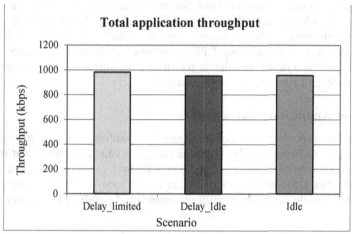

Figure A-4: Downlink overall throughput

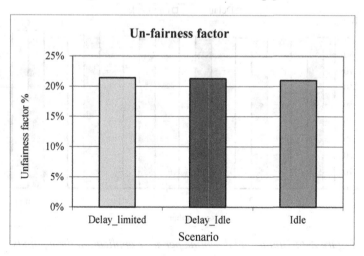

Figure A-5: Downlink overall fairness

A.2   Effects of DSL Fast Mode on HSPA Performance

This investigation is performed to analyze the effect of burst errors on the performance of the HSPA networks. A trace taken from the DSL fast mode operation is deployed in the simulator as described in chapter 2.3.2

The DSL fast mode operation simulation results are compared with the default configuration which is described in Appendix A.1. The DSL default configuration uses random errors without considering the effects of impulse noise and impairment due to neighboring CPEs (Customer Premises Equipments). Both "default" and "fast mode" simulation scenarios are configured with the same FTP traffic model and the same CC configuration.

Per-User Application Throughput

The per-user throughputs for the uplink and the downlink are shown in Figure A-6 and Figure A-7 respectively. From the figures, it can be observed that there is no significant change in per-user throughputs for both configurations. The results summaries that DSL burst errors which are mostly recovered by RLC layers over retransmissions do not have a significant impact on the HSPA performance.

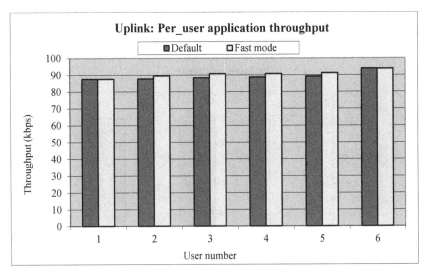

Figure A-6: Uplink and downlink per-user application throughputs

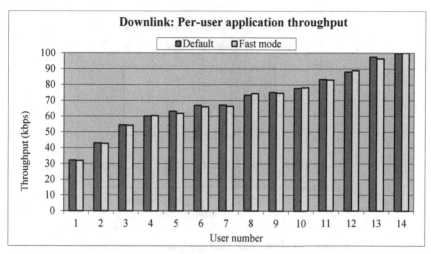

Figure A-7: Uplink and downlink per-user application throughputs

Total Application Throughput

The uplink and the downlink total application throughputs are shown in Figure A-8 and Figure A-9 respectively. As corresponding to the previous argument, both configurations show the same performance for the overall application throughput.

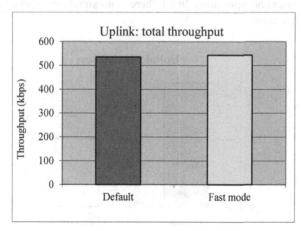

Figure A-8: Uplink total application throughputs

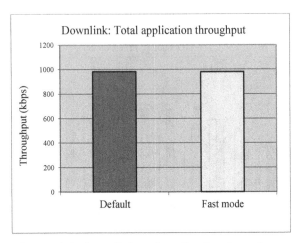

Figure A-9: Downlink total application throughputs

Unfairness Factor

The unfairness factors for the uplink and the downlink are showing in Figure A-10 and Figure A-11 respectively. The uplink fast mode shows better fairness compare to the default mode. However, the unfairness values are not significant and they both show a very good fairness. The fairness factors for the downlink are equal for both configurations. In all, there is no significant impact on fairness for both configurations.

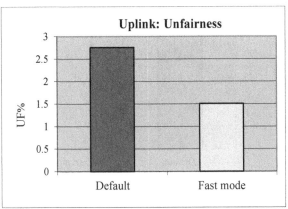

Figure A-10: Uplink total unfairness factor

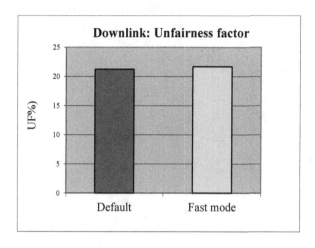

Figure A-11: Downlink total unfairness factor

Conclusion

During this investigation, effects of the DSL transport for HSPA end user performance have been studied. Different QoS parameters such as BER, delay and jitter have been analyzed for the two modes of DSL operations: DSL default mode and fast mode. All DSL based packet errors can easily be recovered by the RLC layer without having impact on the TCP performance. Therefore, simulation results confirm that neither DSL random bit errors nor burst errors have a significant impact on the HSPA performance. Further analysis also confirms that the DSL delay and the jitter not affect the overall HSPA performance. In summary, the limited DSL network based simulations show similar performance which was achieved by the ATM based transport based simulations and therefore the expensive ATM transport links can be replaced by the cheap DSL links without having a severe impact on HSPA performance.

Appendix-B.   LTE Network Simulator

The main objective of the development of the LTE system simulator is to analyze the end-to-end performance and the S1/X2 interface performance. The LTE network simulator is implemented in the OPNET simulation environment. By the time of LTE system network development, there were no LTE based OPNET models. Therefore, the simulator development was started with the basic OPNET modeler modules which consist of standard protocols and network models. As in the HSPA simulator, the LTE network simulator is also used for some of the standard node entities; it modifies them according to the LTE protocol architecture. In this chapter, the LTE system simulator design and development are discussed in detail.

B.1   LTE Reference Architecture

To achieve the aforementioned objectives, the LTE reference network architecture which is shown in Figure B-1 is chosen for the LTE simulator development within the focus of this investigation and analysis. The LTE transport network of the reference model consists of three routers (R1, R2 and R3) in the reference architecture. There are four eNBs that are connected with the transport network via last mile links. EPC user plane and EPC control plane network entities are represented by the Access Gateway (aGW) network entity. The aGW includes the functionalities of the eGSN-C (evolved SGSN-C) and eGSN-U (evolved SGSN-U). The remote node represents an Internet server or any other Internet node which provides one of the end node representations. The average delay for the data transmission between the Internet and aGW is assumed to be approximately 20 ms. Since the network between these entities is mainly wired, a lower delay and negligible packet loss ratio can be experienced.

The transport network between aGW and eNBs is represented by the network which includes the three routers: R1, R2 and R3. This network is not owned by the Mobile Network Operators (MNOs). Therefore, MNOs do not have complete control over these router functionalities. This is an important issue which should be considered at the packet level service differentiations. Mostly MNOs have to control service differentiation for packetized data traffic mainly at the entrance point of the transport network, which means service differentiation should be applied at the aGW for the downlink and at the eNB for the uplink.

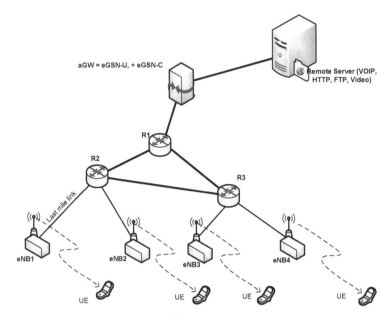

Figure B-1: LTE reference architecture

However, for the downlink direction, the last mile is the bottleneck which is located between the eNB and the end router, the router which is closest to the eNB. The transport network DL and UL data flows are shown in Figure B-2.

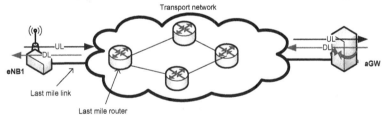

Figure B-2: LTE transport network DL and UL data flows

Any overload situation at the last mile link causes packet losses mostly at the end router which is directly connected with the last mile link. Therefore, a certain degree of service level guarantees also have to be provided by the routers in the transport network. Since the last mile in the uplink is directly located at eNB, traffic differentiation can be more effectively controlled at eNB by MNOs for the uplink transmission.

Inside the reference LTE architecture used in this investigation, there are four eNBs where each of them controls three cells. As mentioned in chapter 2.4, the eNB is one of the key elements in the E-UTRAN network. It connects to the transport network through the S1 and X2 interfaces at one side of the network; at the other end, it connects with users through the Uu interface. Therefore it includes all radio interface related protocols such as PHY, MAC, RLC, PDCP etc. and all transport protocols such as GTP, UDP, IP, Layer 2 and Layer 1 (Figure 2-15). At the transport side of the eNB it provides service differentiation for the uplink data flow when it sends data to the transport network. Any congestion situation at the last mile has to be controlled by eNB for the uplink and therefore the uplink service differentiation along with traffic shaping plays a significant role for data handling. These aspects of eNB are discussed in detail in this chapter.

B.2   LTE Model Architecture

Based on the above reference protocol architecture, Figure B-3 shows a simulator overview of the LTE system simulator in the OPNET environment. Each eNB can be configured with a variable numbers of users. For example, the proposed simulator overview shows a 10 UEs/cell configuration however simulation model is flexible of configuring any user constellation along with the mobility based on requirements of the analysis. Required applications for each user and also the servers are configured in the application configuration (upper left corner) and user configuration profiles. In OPNET, profiles group the users according to their selected application these configured user profiles have to be activated at each UE entity.

Any mobility model can be included within the mobility network and can activate any user based on their individual mobility requirements. Currently most generic mobility models such as Random Way Point (RWP) and Random Directional (RD) are implemented and readily available for the configuration. As mentioned above, all UE mobility information is stored and updated regularly in the central database (also called global user list) where it can be accessed at any time.

Three main node models have been developed with respective user plane protocols within the simulator. They are

- UE node model,
- eNB node model,
- aGW node model.

Next, the details about these node models are discussed.

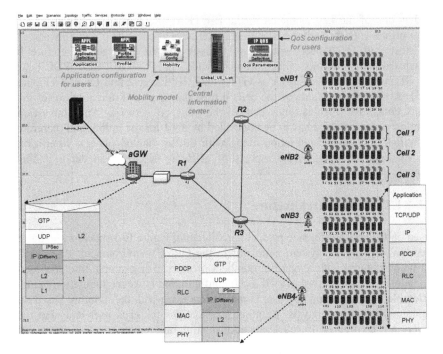

Figure B-3: LTE simulation model overview

UE Node Model

The UE node model is created according to the layered hierarchy shown in Figure B-4. When creating a UE in the network simulator, one instance of this node model is created. All the attributes related to the layers can be configured individually. The node model includes two sets of protocols: the LTE air interface related protocols and the end user application protocols. The LTE Uu protocols include PDCP, RLC, MAC and PHY layers. Except for the PHY layer protocol, all other protocols have been implemented according to the 3GPP specification [47] in the respective layers. The physical layer is modeled to transmit the transport blocks of user data between eNB and UE entities in order to model physical layer characteristics from the system simulator point of view without having explicit detailed modeling. This is done by the MAC scheduler which has been designed to emulate the PHY layer characteristics in terms of data rate or TB size/TTI for each UE entity. More details about the MAC scheduler and the MAC layer protocol functionalities are discussed in chapter 2.4.3.

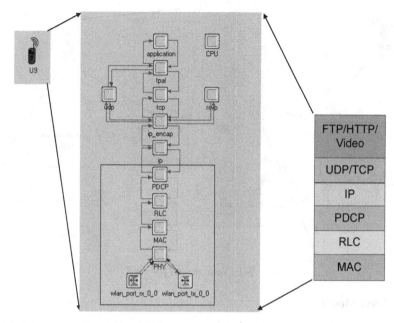

Figure B-4: Node model for UE

eNB Node Model

The protocol layers of the eNB node model are shown in Figure B-5. The creation of any eNB includes the instance of this node model. It includes all the protocols which are required by the standard eNB entity. The attributes of each layer in the eNB can be configured independently. The right hand side protocols of the node model in figure B-5, the PDCP, RLC and MAC layers represent the Uu related functionalities whereas the left hand side protocols of the node model, the GTP, UDP, IP and Ethernet layers represent the transport related functionalities. The MAC scheduler is located at the MAC layer in the eNB. It is the key functional element which interacts with all other Uu protocols. There are three schedulers which are implemented on a cell basis within a single eNB node model. Each scheduler works independently based on the cell configuration. The core controller which connects the data flows from three cells combines and forwards the data flows to the upper layers. IPSec and the service differentiation are performed at the IP layer in the eNB. The diffserv model is implemented in the eNB node model mainly for the uplink and partly for the downlink. The model includes traffic differentiation and scheduling in the IP layer and shaping functionalities at the Ethernet layer. More details about those functionalities are given in the chapter 2.4.3.2.

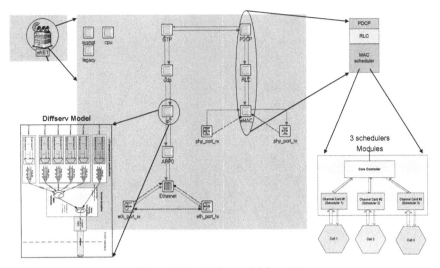

Figure B-5: Node model for eNB

aGW Node Model

aGW user plane transport protocols which are implemented in the simulator are shown in Figure B-6. One part of the peer to peer transport protocols, GTP, UDP, IP and Ethernet, are at the side of the transport network whereas the other part is at the side of the remote server side. Figure B-6 shows the two GTP process models, GTP0 and GTP1, which can be used for the uplink and downlink separately if two end nodes are connected for those traffic flows. If one server is used for the downlink and the uplink applications then either GTP0 or GTP1 can be used for both functionalities. In addition to the diffserv model, the IPSec functionality (according to specification [46]) is also implemented within the IP layer in both aGW and eNB node entities in order to provide data security over the transport network. There are two Ethernet interfaces which can be linked to the transport network based on the connectivity. Currently one interface is connected to the transport network. The other interface is left for any other extension of the model further investigations.

So far, an overview about the LTE system simulator has been discussed. During the LTE simulator development, most of the standard protocols such as application, TCP, UDP and all adaptation layer protocols were used without any modifications. However some of the protocols such as IP and Ethernet are modified according to the LTE requirements. There is another set of protocols which is newly added to the simulator based on the specific requirement of the 3GPP LTE network and it has been discussed in chapter 2.4.3. The next section discusses the modified and new protocols in detail. Further it provides an overall protocol user plane architecture which is the main focus of the simulator development.

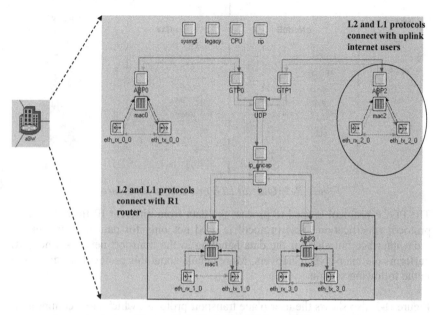

Figure B-6: Node model for aGW

B.3 LTE Protocol Development

Figure B-6 shows the overall LTE user plane protocol architecture which is used to develop the LTE simulator. All protocols can be categorized into three groups, radio (Uu), transport and end user protocols. The radio (Uu) protocols include the peer-to-peer protocols such as PDCP, RLC, MAC and PHY between the UE and eNB entity. The detailed physical layer modeling is not considered in this simulator. However, effects of radio channels and PHY characteristics are modeled at the MAC layer in terms of the data rates of individual physical channel performance. The modeling of PHY layer characteristics within the MAC layer is discussed later in this section.

The RLC is used to recover any losses over the air interface which cannot be recovered by the MAC HARQ mechanism. The RLC protocol is implemented in the simulator according to the 3GPP specification [73]. Since the LTE RLC protocol functionalities are not different from the HSPA implementation, the detailed implementation procedure is not discussed in this chapter.

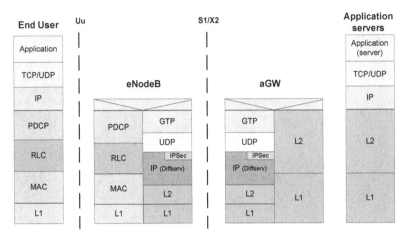

Figure B-7: Overall LTE protocol architecture

The PDCP protocol is used to process the data from different IP flows based on protocol specification. This protocol is used not only for data handling in the radio interface but also for the data handling in the transport network when UE performs the inter-eNB handovers. More details about this protocol are discussed in the following section.

Figure B-7 also shows the user plane transport protocols which are a common set of protocols that are used in both the S1 and X2 interface. It includes the GTP, UDP, IP and L2 protocols. Ethernet is used as the layer 2 protocol for the current implementation; however the model is implemented in such a way that it can use any other transport protocol with a slight modification to the node model. IP is one of the key protocols which handle routing, security (IPSec), service differentiation and scheduling functionalities at the transport network. The default OPNET IP layer is largely modified by adding the above mentioned functionalities IPSec and diffserv based on the specific LTE requirements [49].

All control-plane protocol development is not directly considered within the LTE simulator. However effects of the signaling such as overhead and delay are taken into consideration at the respective design of the user-plane protocols.

PDCP Protocol Development

The PDCP layer user-plane protocol is also implemented according to the 3GPP specification [49] in the UE and the eNB entities. The state process model of the PDCP protocol implementation is shown in Figure B-8.

The main PDCP functionalities which are implemented in this protocol can be summarized as follows:

- bearer identification and packet classification based on the traffic QoS,
- per-bearer based PDCP buffer management,
- encapsulation and de-capsulation of PDCP PDUs and packet forwarding techniques,
- data handling during inter-eNB handovers (X2 packet forwarding, reordering and error control etc.).

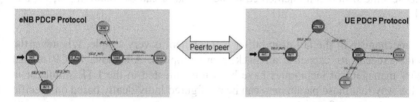

Figure B-8: Peer-to-peer process model for PDCP protocol

After setting up initial bearers, based on IP packet arrivals, each bearer connection is identified and classified using IP header information. Each bearer connection has individual PDCP buffers and data is stored in these buffers until it is served.

For the LTE network, the main data buffers are located at the PDCP layer. The PDCP buffers have been implemented on a per-bearer basis. However they use common shared memory space in the eNB entity. In contrary to this, at the UE side, the shared buffer capacity is allocated on a per-user basis. The timer based discard, RED (Random Early Detection) and the simple drop-tail buffer management techniques have been implemented on these per-bearer PDCP buffers and are explained in the following.

The timer based discard mechanism discards the packets if they exceed the configured maximum delay limit. Packets can be delayed for a long time mostly due to two reasons: either the radio channel of the user is very bad or some TCP connections staled or timed-out due to congestion in the network. In the first case, since the user channel quality is bad, by discarding the packet at PDCP, feedback is provided to the TCP source for flow control and hence the offered user load is controlled by TCP dependent on the available channel capacity. In the second case, packets due to an invalid TCP connection should be discarded before they are being transferred over the Uu interface. Not only for the best effort traffic sources but also for real time traffic sources, such a delay-based discard mechanism can be beneficial. It also avoids unnecessary transmissions of delayed packets over the network. In a nutshell, this buffer management technique avoids wastage of the scarce radio resources and enhances network resource utilization.

The RED buffer management scheme uses a TCP friendly random discarding approach. It maintains a lower and an upper buffer limits with two different drop probabilities. Based on the filling level of the buffer from lower threshold to upper threshold, the packet discarding probability increases linearly. Due to the random discards, in-sequence multiple losses are avoided and TCP adapts to the data flows dependent on the capacity of the network. Further details about the RED scheme which is implemented in the simulator can be found in [74, 75, and 76].

The mechanism of the tail-drop buffer management is very simple. Whenever the buffer exceeds its capacity, all packets are dropped. All above PDCP related buffer management techniques have been implemented in the LTE simulator in such a way that these parameters can be configured individually based on specific QoS requirements for each case.

Apart from buffer management techniques, all other PDCP related protocol functionalities such as data encapsulation, decapsulation, forwarding etc. have been implemented according to the 3GPP specification [49]. Before sending the packet to the lower layer, the PDCP layer encapsulates the IP packets into PDCP PDUs. In contrary to this, the decapsulation of data from PDCP PDUs is performed at the receiver PDCP entity. The PDCP protocol is also responsible for secure X2 forwarding during handovers. Mainly, the implementation supports the reordering, in-sequence delivery and error handling (ARQ) during inter-eNB handovers.

GTP and UDP Protocol Development

At the transport side, GTP (GPRS Tunneling Protocol) creates the transport tunnel over the UDP protocol between eNB and other end-node which can be either aGW or peer eNB entity. The state model of the GTP protocol is shown in Figure B-9. At the aGW side, there are two GTP state models GTP1 and GTP2 running in parallel which process the uplink and downlink flows separately when flows are distinct to different end nodes. The same processes model is used for both nodes. When the aGW is only connected to a single server, one process, either GTP0 or GTP1, can be used for both flows. All these GTP protocol functionalities have been implemented in the simulator according to the 3GPP specification (46, 49).

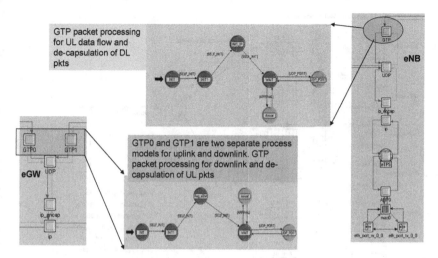

Figure B-9: GTP protocol process model

IP Diffserv Model Implementation

In addition to the generic IP routing protocol functionalities, the advanced service differentiation model (diffserv) is implemented in the IP layer of the eNB for the uplink. The diffserv model consists of three main functional units, which are the packet classification unit, the scheduler unit and the buffer management unit. According to the QoS requirements, the service flows are differentiated and packets are classified using the Differentiated Services Code Point (DSCP) values on the diffserv field (type of services, ToS field) in the IP header of the IP PDU. The LTE transport network uses the differentiated service architecture for transport network packet classification which is defined in RFC 2474 and RFC 2475. The DSCP values are used to mark the per-hop behavior (PHB) decisions about the packet classification to provide the appropriate QoS treatment at the transport network. After the packet identification and classification process they are stored in the respective PHB (Per-Hop Behavior) buffers. The complete diffserv model with PHBs is shown in Figure B-10.

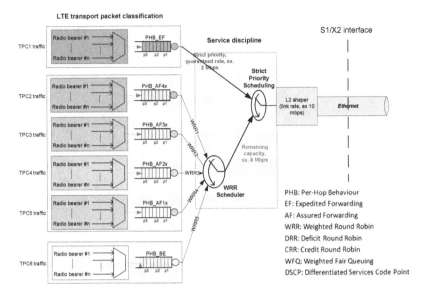

Figure B-10: Diffserv model architecture

The PHBs can also be categorized into three distinct QoS classes, which are Expedited Forwarding (EF), Assured Forwarding (AF) and Best effort (BE) or default PHB. The LTE uplink diffserv model is built with 1 EF PHB, 4 AF PHBs and 1 BE PHB which are used to classify the incoming service flows at the IP layer. Three virtual priority queues (drop precedence) are created within each AF PHB physical queue which further prioritizes the service flows based on their QoS requirements. The detailed classification which is implemented in the diffserv model is shown in Table B-1.

Table B-1: Assured forwarding service classification

Drop precedence	Assured Forwarding (AF) PHBs			
	PHB_AF4x	**PHB_AF3x**	**PHB_AF2x**	**PHB_AF1x**
DP Level 1	PHB_AF41 (DSCP 34)	PHB_AF31 (DSCP 24)	PHB_AF21 (DSCP 18)	PHB_AF11 (DSCP 10)
DP Level 2	PHB_AF42 (DSCP 36)	PHB_AF32 (DSCP 28)	PHB_AF22 (DSCP 20)	PHB_AF12 (DSCP 12)
DP Level 3	PHB_AF43 (DSCP 38)	PHB_AF33 (DSCP 30)	PHB_AF23 (DSCP 22)	PHB_AF13 (DSCP 14)

The scheduler unit has been implemented to provide the required service priorities for these PHBs. It consists of strict priority queuing and weighted fair queuing disciplines. Since EF PHB handles the services which are delay, loss and jitter sensitive such as signaling and control traffic, the scheduler provides strict priority for this PHB class Furthermore, it provides serving capacities for the AF classes based on their configured weights. In the weighted fair queuing scheduler, the lowest priority is assigned to a BE PHB class which is used to handle non-real time services such as HTTP traffic and FTP traffic. Whenever the AF PHBs are not using the capacity based on their assigned weights, the remaining excess capacity is always allocated to the default (BE) PHBs.

Except EF PHB, all others use the RED or the weighted RED (WRED) buffer management techniques which are implemented in the BE PHB and the AF PHBs respectively [77]. The WRED combines the capabilities of the RED algorithm to provide preferential traffic handling of higher priority packets. It selectively discards lower priority traffic when the interface begins getting congested and provides differentiated performance characteristics for different classes of services. More details about the implementation of RED and WRED is given in the reference [75, 76].

Ethernet Shaper Implementation

The L2 shaping functionality is implemented at the Ethernet layer and provides serving triggers to the diffserv scheduler. The shaping model is a single rate shaper which uses the token bucket principle [50]. The configurable CIR (Committed Information Rate) rates and the CBS (Committed Burst Size) are the two main configurable parameters for the shaper. The basic shaper functionalities are shown in Figure B-11.

The shaper receives Ethernet frames from its input buffer, which are formed as soon as IP packets are picked from the scheduler. The shaper's input buffer is treated as a FIFO, and its size is set to a configurable value (for example 4 KB). In order to optimize the transport network utilization, the shaper generates a scheduling request towards the diffserv scheduler whenever the input buffer has at least a certain empty buffer space, for example 2 KB. Based on the CIR, arriving Ethernet frames are served to the lower layer using a token bucket scheme which is described in [50].

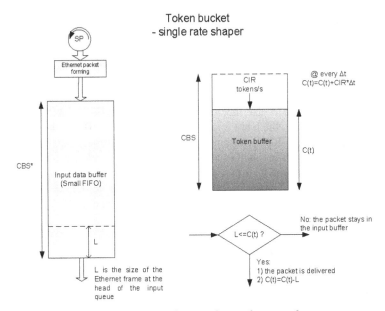

Figure B-11: Ethernet shaper functionality

LTE MAC Scheduler

The LTE MAC scheduler is one of key elements located in the eNB entity. It distributes the available radio resources among the active users. As discussed in chapter 2.4, it operates in both time and frequency domain by sharing the resources in small units called PRBs (Physical Resource Block). The MAC scheduler is directly connected with the eNB Uu protocols such PDCP, RLC and PHY through the signaling channels. All these information is used in order to take decisions when allocating resources to the users in an effective manner. This section summarizes the eNB scheduler design and implementation for the LTE system simulator.

The eNB maintains three cells, where each of them has two independent MAC schedulers for the uplink and the downlink. The state process model for the scheduler is shown in the following Figure B-12. It consists of 6 decision state models and one arrival state model. All other states are used for the initialization and signaling activities.

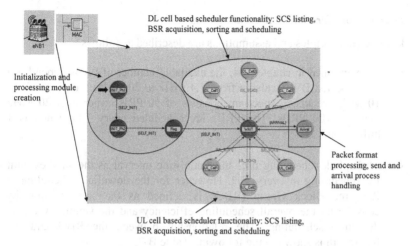

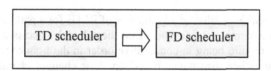

Figure B-12: MAC scheduler process model for eNB

From this point onward, the common platform for the uplink and the downlink MAC schedulers is discussed in detail in the following paragraphs. Differences between the uplink and the downlink scheduling are highlighted separately.

From the design and implementation point of view, the scheduler consists of a time domain (TD) scheduler and a frequency domain (FD) scheduler which are shown in Figure B-13.

Figure B-13: Architecture of the eNB MAC scheduler

The TD scheduler performs the user identification, QoS classification and bearer prioritization as the main functions. The FD scheduler effectively serves available radio resources among the users who have been already prioritized and selected by the TD scheduler. In this case, depending on the scheduling disciplines the amounts of resources which are allocated to each user are served by the FD scheduler. For the current LTE simulator, two main scheduling disciplines have been implemented in the LTE simulators which are fair scheduling (Round Robin, RR) and exhaustive scheduling.

Key Design Considerations and Assumptions

The key design principles and assumptions are described as follows.

- As mentioned in chapter 2.4, the available number of PRBs depends on the allocated BW (scalable from 1.4 MHz to 20 MHz). For example, a 10 MHz frequency spectrum consists of 50 PRBs whereas a 20 MHz includes 100 PRBs. The PRBs are scheduled every TTI which is 1 millisecond.

- The uplink scheduling uses the same time interval as the smallest unit for scheduling in every TTI. However for the downlink scheduling, a Resource Block Group (RBG) is defined as the smallest unit by considering the overall scheduling efficiency and the simplicity of the downlink scheduling. The relationship between the RBG and the bandwidth is shown in the following Table B-2.

Table B-2: LTE supported BWs

LTE supported BWs, PRBs and RBGs						
BW (MHz)	1.4	3	5	10	15	20
#PRBs	6	15	25	50	75	100
#PRBs per RBG	1	2	2	3	4	4

- It is possible to allocate any number of RBGs for a particular connection which is completely dependent on the scheduler decision. These RBGs do not require being in consecutive order in the frequency domain; they can be rather distributed based on their channel characteristics with different frequencies and MCS. However all PRBs within one RBG use the same MCS values to achieve a high processing gain by reducing the complexity. For the uplink scheduling, individual PRBs are allocated to users and they are required to be in consecutive order in the frequency domain.

- The signaling (SRB1, SRB2 and QCI5) has low data volume compared to the offered traffic volume and the signaling packets are also not as frequent during the transmission as other traffic. Therefore the signaling overhead is neglected for the scheduling which is one of the key assumptions for this scheduler design.

- Downlink scheduling is done on a per-bearer basis whereas uplink scheduling is done on a per-connection or per-UE basis. Further the first phase of the scheduler design only considers a few QCI bearer types such as QCI1, QCI7, QCI8 and QCI9.

Procedure for Scheduling

The scheduling procedure can be described by three main steps. They are

- bearer identification and bearer classification,
- bearer prioritization and resource allocation,
- Scheduling and resource assignments.

The first two steps (a and b) mainly describe the TD scheduler functionality which considers the QoS aspects and buffer occupancies for each active bearer connection whereas the last step focuses on the FD scheduler functionality which defines the physical layer characteristics of the bearer connection along with the resource allocation in terms of PRBs.

Bearer Identification and Bearer Classification

The Scheduling Candidate List (SCL) is created by listing all active users who have data in their buffers. The BSR (buffer status report) provides the details about the buffer occupancy for each bearer connection which indicates the data availability at PDCP and RLC transmission and retransmission buffers including their volume information. For the system simulator, it is assumed that the BSR reports are received in the MAC entity with zero delay. Depending on the QoS priorities and the pending retransmissions, the selected bearers which are in the SCL list are grouped. For example, separate lists can be defined based on this classification such as SCL-1 for pending retransmission bearers, and SCL-2 for data bearers.

Bearer Prioritization and Resource Allocation

All users who have pending retransmissions are selected with highest priority for the scheduling. All others are prioritized based on a common weighting factor which is defined based on several factors such as the bearer type either GBR or non-GBR, channel characteristics, buffer occupancy and the accumulated data volume. This common weighting factor can be divided into two main categories, the Proportional Fair (PF) term and the bearer QoS type (GBR or Non-GBR) based weighting term. For the LTE system simulator, the current channel capacity of the bearer is defined by the PF term. The GBR-related weight factor is defined for the real time services such as the VoIP whereas the Non-GBR related weight factor is defined for the best effort services such as web browsing and large file downloads. Finally the simplified version of the common weight factor (CWF) for scheduling is defined as follows.

$$CWF = PF \cdot (WF_{GBR\ related} + WF_{Non-GBR\ related})$$ *equation 6-1*

Based on this common weight factor rating, the users in the data bearer list are prioritized for the scheduling. The PF term includes the per-UE throughput which is calculated based on two key parameters, which are the number of PRBs (or RBGs) and the final MCS value for a particular connection. The input parameters such as MCS can be taken either from literature or dedicated radio simulation. The number of PRBs is selected by the FD scheduler based on the selected scheduler discipline which is discussed in the next section.

Scheduling and Resource Assignment

Two FD schedulers, the Round Robin (RR) and the exhaustive scheduler are implemented in the simulator. The RR FD scheduler uses a fair approach compared to the exhaustive FD scheduler. Both schedulers use maximum up to "n" UEs for scheduling in each TTI. If the number of users who have data in the transmission buffer is less than the maximum number ("n"), then these users are taken for the current scheduling, otherwise always "n" UEs are scheduled in a TTI.

For the RR scheduler, the number of PRBs for each user is calculated simply by dividing the spectrum by the number of users, "n".

$$number\ of\ PRBs\ per\ UE\ = \frac{PRBs_{spectrum}}{n}$$ *equation 6-2*

The number of PRBs for DL transmission should be determined in terms of RBGs which depends on the configured BW as described at the beginning of the scheduler description. Depending on the channel, users can achieve different data rates with the same number of PRBs for the RR FD scheduler.

The exhaustive FD scheduler in UL and DL allocates full resources to the UE who is at the top of the priority list for the current TTI. The priority list is created by the TD scheduler. The maximum resource limit of the UL priority user is decided either by UE power limitation and also by the input taken from the radio simulator or the available data volume for the exhaustive FD scheduler whereas the DL users are allowed to use even full spectrum within the current TTI if they have sufficient data in the buffer for the current transmission. When the MCS and the number of PRBs are known for each bearer, the transport block size (TBS) is determined using table 36.213 of the 3GPP specifications.

Modeling HARQ

For modeling HARQ, a simplified approach is used as in HSPA. It can be assumed that the BLER is 10% with a maximum of three retransmissions. The 2nd and 3rd retransmission have the effective BLER of 1% and 0% respectively. The configurable interval between two consecutive retransmissions is 8 ms for this investigation. Table B-3 below shows the BLER with the number of retransmissions.

Table B-3: BLER with the number of retransmissions for HARQ

	1st transmission	2nd transmission	3rd transmission
Effective BLER	10 %	1 %	0 %

A uniformly distributed random number is used to model above BLER and retransmissions. If the uniformly distributed number is greater than the chosen BLER, then the current transmission is successful, otherwise it is unsuccessful. If the first transmission is unsuccessful, the packet (TBS) and related information (e.g., the number of PRBs) are stored in a separate buffer and a list respectively. A retransmission timer is set up for the next trigger. After 8 ms, at the expiry of the retransmission timer, the second retransmission attempt is triggered for the previous packet again by using the same probabilistic approach as for the first retransmission. If the second retransmission still fails, then again after 8ms, the original packet is sent as the final successful transmission (0% BLER). All these three retransmission attempts use the same number of PRBs. All due retransmissions have the highest priority for scheduling in the current TTI.

B.4   LTE Mobility and Handover Modeling

A mobility model represents the realistic movements of the mobile users. There are several mobility models which describe different mobile user behaviors [51]. Some of them represent independent user mobility in which user movements are completely independent on each other whereas some others represent dependent user movements [51]. When it comes to broadband mobile networks, the mobility of the user plays a key role for signaling and traffic load handling. A large number of mobile users who are frequently changing cells and different networks lead to the creation of a huge signaling and data handling amount across the mobile network where delay requirements have to be strictly kept. Therefore, an accurate modeling of such user mobility is very important for dimensioning of the mobile broadband network.

Within the LTE system simulator, mainly UE mobility is considered for the intra-LTE (intra-eNB and inter-eNB) handover modeling. The details of intra-LTE handovers have been discussed in chapter 2.4.5. The main goals of this analysis are to evaluate the impact of the additional load due to handovers on the transport network (X2/S1) performance and on the end user performance. Further, in order to provide seamless mobility for end user, the handover data must not be delayed over the transport network and is required to be prioritized. The intra-LTE handovers along with the user mobility are modeled within the simulator to investigate and analyze all aforementioned objectives. The individual handover frequency of each connection and its carried load are the key parameters that have to be considered during handover modeling. The UE handover frequency depends on many factors such as the selected mobility model, the UE speed etc. Depending on the number of users in the cell and their traffic models, the offered handover load over the transport network also differs. Therefore by considering all these factors, a simple, novel approach for intra-LTE handovers is designed and implemented in the LTE simulator. The new approach provides a great flexibility to use any mobility model or to configure any handover scenario individually based on the focused investigation.

Within the new approach, the UE mobility which mainly provides the carried data rates during handovers depending on the location of the UE (or distance to eNB) and the UE handover decision are decoupled. Therefore, the model provides a great flexibility to use any mobility model. Further the handover frequency can be configured separately within the model and hence different handover scenarios can be configured to analyze the impact on the transport network individually upon the focused investigation.

The section is organized as follows. A brief overview about generic mobility models and the design and development of an enhanced mobility model are described in the following paragraphs. A modified random directional mobility model is developed according to the reference architecture which was described in Appendix B.1.

This model provides the UE location details to the MAC scheduler in the serving eNB to manage the resources among the active UEs in the cell. A novel approach of modeling intra-LTE handovers and X2 data management for inter-eNB handovers are described at the end of this section.

Generic Mobility Models

The commonly used generic mobility models are the random walk [51, 52], random way point and random direction [51]. All these three models use the random mobility of mobile nodes within a predefined area. The speed and direction changes occur at a reasonable time slot and the different movement parameters are altered within predefined ranges, for example the speed ranges between maximum and minimum values and the direction within the range 0 and 2π. The main difference between the random walk and the random way point mobility models is that the random walk mobility model does not use a pause when a user changes his direction of travelling as in the random way point mobility model. However these two mobility models have the issue of a "density wave" in which the most of the mobile nodes are clustered to the center of the area. To avoid such clustering effects and also to ensure that all mobile nodes (MNs) are randomly distributed over the area, the random direction mobility model is introduced. In this mobility model, the mobile nodes or users change the direction of travelling at the boundaries of the predefined area. The mobile node travels towards a selected direction with a certain speed until it reaches the boundary of the predefined area and then changes the direction randomly after a certain pause period. By studying all these generic random mobility models, an enhanced mobility model is developed for the LTE system simulator. Further the inter-LTE handovers are also implemented within this mobility model.

Enhanced Random Mobility Model

As described in chapter 2.4.5, the LTE system simulator is configured with 4 eNBs running 3 cells each for this investigation. Therefore users who move among these 4 eNBs have inter-eNB and intra-eNB handovers.

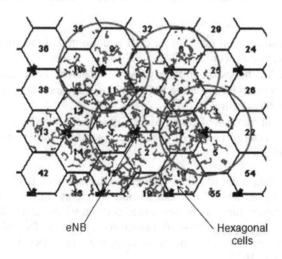

Figure B-14: Cell architecture of four eNB scenario

Figure B-14 shows the generic view of the cell architecture and user mobility within those cells and eNBs for a certain scenario. For the modeling of the mobility within the reference LTE architecture, the 4 eNB along with 3 cells each are selected as marked in the Figure B-14. Each eNB maintains three hexagonally shaped cells. Four circles are drawn to represent the coverage of the four eNBs inside the LTE simulator. Users randomly move from one cell to another by making inter-cell and inter-eNB handovers which mostly occur at the edges of the cell coverage. Intra-eNB handovers do not have a significant impact on the transport network since the main data buffering occurs at the same eNB. However the situation is more crucial when a mobile node is performing inter-eNB handovers. As described in chapter 2.4.5.1, the buffered user data in one eNB has to be transferred to the other eNB where the mobile node is moving to. All this data management within the transport network is discussed in this section.

The simplified mobility modeling is considered to implement user mobility when a UE performs intra-eNB and inter-eNB handovers within the reference LTE architecture. Mobile nodes are moving according to the modified random directional mobility model across the circular coverage area of the eNB. Here, modified random directional mobility means whenever a mobile node changes the direction by reaching the boundary, it does not stop moving, i.e. the pausing period is zero as in the random walk mobility model. Since the handover frequency does not depend on the user mobility for the new approach, the following key assumptions are taken for the development of the LTE mobility model.

- The transmission range of an eNB is uniform and the coverage area is represented by a circle.
- All mobile nodes are uniformly distributed across the coverage area.
- Only four eNBs are considered for this implementation and one eNB covers three cells.
- The total number of active users within these 4 eNBs is constant. All users perform handovers and user mobility within this four eNB constellation.

From the modeling point of view, each eNB coverage area is divided into two zones denoted by an inner circle and an outer circle as shown in Figure B-15. Inter-eNB handovers can only be possible to perform when the mobile terminal is in outer zone. However intra-eNB handovers can be performed within both circles. The three cells which form the coverage area are modeled by three sectors as shown in Figure B-15.

The next section describes the statistical approach for the intra-LTE handovers in which handover decisions are taken for each connection independently.

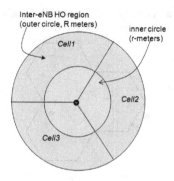

Figure B-15: Coverage area zones of a eNB

Statistical Approach for Intra-LTE Handovers

The handover decision is taken according to the statistical approach that can be taken from the literature for each mobile node separately. For example, a UE handover occurs uniformly distributed between "t1" and "t2" with mean "tm". In this case any probability distribution can be deployed for the handover decision of each Node-Based on the probabilistic distribution or trace. Since simulations are performed for a considerably long period of time within the system simulator, the statistical measures of intra LTE handovers such as handover frequency in terms of mean and variance are important for each mobile node. In such cases, the exact load over the transport network can be simulated based on their traffic model. According to above example scenario, the mobile node can take a handover decision to any of the 4 eNBs with the mean interval of "tm". However when the handover decision is taken, the UE may be outside of the outer circle in the eNB. In this case, the UE cannot make the handover decision immediately and has to hold the decision for example "Δt" time interval until it reaches the outer circle of the eNB. Therefore, the mean inter-eNB handover duration should be calculated according to the following equation.

$$t_m = t_{dist} + \Delta t$$

This factor depends on the speed of the UE. For faster speeds this time is very low

t_m : is the expected mean, for example 45 sec

t_{dist} : is mean of random distribution. for example uniform (40 sec)

Δt : is the time required for UE to move from inner circle to outer circle according to the current travelling direction. when UE is in outer circle, Δt becomes zero.

equation 6-3

If the handover decision occurs when the mobile node or UE is in the outer zone of the eNB, then "Δt" is equal to zero and the handover is performed immediately. The general concept of inter-eNB handovers can be described according to the following example Figure B-16.

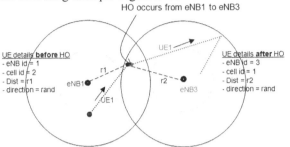

Figure B-16: Inter-eNB handovers functionality

In this example scenario, the UE is moving from eNB1 to eNB3 and the handover occurs within the outer circle area. The distance to eNB1 just before the HO is represented by r1 and just after the handover (to eNB3) by r2. During the inter-eNB handover process the parameters eNB_id, cell_id, distance to transmitter and direction (eNB3 point of view) is changed.

Since there are four eNBs in the simulator, the UE can randomly perform inter-eNB handovers to any of the other eNBs. Therefore the eNB id can be randomly selected one out of the numbers {1, 2, 3 and 4}. The new cell id can also be selected one out of the number {1, 2 and 3}. The distance to the transmitter (or target eNB) can be any point in the outer zone and also the direction can also be random. By considering all above facts, the modeling of inter-eNB handovers is simple and efficient. Whenever the handovers occur, these mobility parameters can be selected for each node separately which represents the real handover situation from the system simulator point of view. The important fact for the transport network in order to perform X2 forwarding is how often the inter-eNB handovers occur as well as the timing and the changes of eNBs. This example handover scenario is also elaborated in Figure B-17 with one eNB convergence zone.

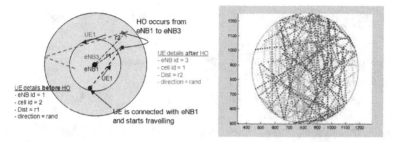

Figure B-17: Inter-eNB handovers mobility representation

Further, Figure B-17 on the right shows the four users' mobility with inter-eNB handovers within the LTE simulator. The four colors represent the four users' mobility behavior separately.

As shown in the above example, the UE parameters are changed at the handover location according to the probabilistic approach. All user information including mobility and location information are sent and stored in the central data base. As soon as the UE makes any change about its location due to a handover, this will be reported to the central database and also are notified to the corresponding network entities which have been affected. Further details about the inter-eNB handover data handling within the transport network is discussed in the next section.

All above discussion is about the inter-eNB handovers; in comparison, the modeling of intra-eNB handovers is simple since it does not require data handling over the transport network. During intra-eNB handovers, all data handling is performed within the eNB and it is also the main controlling unit for handover procedures.

Inter-eNB HO Data Handling

This section mainly focuses on data handling in the transport network during inter-eNB handovers. Within the LTE simulator, the models for the mobility and the transport network are linked through a central data base.

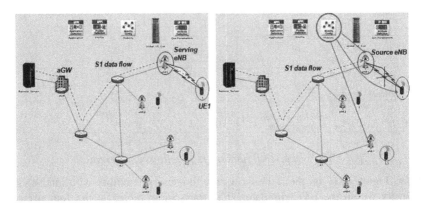

Figure B-18: Inter-eNB handover example scenario-1

As mentioned above, whenever a mobile node starts an inter-eNB handover, the source and target eNB are informed by the mobility network model. The complete inter-eNB handover process can be explained by using the above example scenario. Four UEs numbered 1 to 4 are connected with 4 eNBs in this configuration. In a similar manner, four correspondent mobile nodes are created within the mobility model which is assigned to the corresponding eNB ids and cell ids. The mobile nodes start moving based on the configured speed in the UE configuration with random directions. When a mobile node performs an inter-eNB handover, the following processing is performed for a successful handover.

Figure B-18 shows that the mobile node is currently connected with eNB1 and data to the mobile node is transferred via the S1 interface through the transport network. The UE1 is for example downloading a file from an Internet server where the data flow is originated. The UE1 is directly mapped to the mobile node 1 in the mobility network model. The right side of Figure B-18 shows that UE sends the handover indication (chapter 2.4.5) to both the target eNB and the source eNB. Once the handover starts at the source eNB, it stops sending data to the UE and starts forwarding data to the indicated target eNB. All these events such as handover indication, stopping the transmission and the start of sending data over the X2 to the target eNB occurs based on the predefined time period which was taken from a trace file or from derived distributions. All incoming data is buffered at source eNB and forwarded to the target eNB via the X2 interface while the mobile node completes its handover procedure over the Uu interface (the duration is again determined using a statistical approach or a trace).

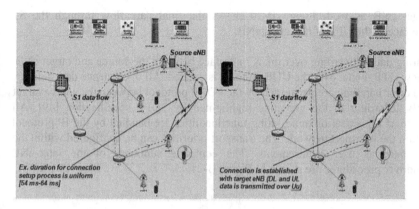

Figure B-19: Inter-eNB handover example scenario-2

Figure B-19 shows this data forwarding process over the transport network between the source and target eNB.

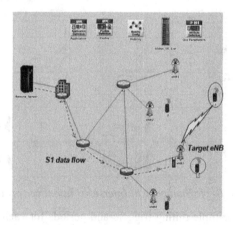

Figure B-20: Inter-eNB handover example scenario-3

Once the UE is successfully connected with the target eNB, all buffered data is immediately transferred to the UE. As discussed in chapter 2.4.5, the target eNB also starts the path switching process by sending a signal to the aGW at the same time. When the path switching process is completed, the aGW switches the S1 data path directly to the target eNB and also triggers the end markers to terminate the data flow through the source eNB path. When the end marker is received by the source eNB, it transparently forwards it to the target eNB by terminating the handover process in its own entity. When the end marker is finally received by the target eNB, all X2 forwarding and handover activities are terminated by the target eNB as well. Figure B-20 shows the final data flow through the S1 interface to the UE1 of the target eNB.

During data forwarding over the X2 interface between the source and target eNB, a new GTP tunnel over the UDP protocol is created. All in-sequence data delivery and error handling are performed by the PDCP protocol between the source target eNB according to the guideline highlighted by 3GPP specification [49]. All transport prioritization and routing functionalities are handled by the IP protocol within the transport network. The transport priorities can be configured within the simulator as user requirements. The complete summary of the inter-eNB handover procedures are shown in Figure B-21.

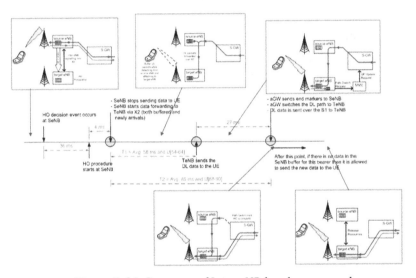

Figure B-21: Summary of Inter-eNB handover procedur

It also shows the example timing triggers and periods for each handover trigger. Further, Figure B-21 shows that the end marker arrival time is dependent on the traffic load situation in the transport network. When the transport is congested, the end marker traverse duration through S1 is longer compared to non-congested transport. Therefore the overall handover duration is determined not only by the statistical time periods but also by the transport network loading situation as well. For the worst case scenarios, there are expiry timers which are activated to terminate the handover process by respective network entities. One example scenario is that if the end marker is lost during the transmission due to the congestion in the transport network, it will not be received by either the source eNB or target eNB. Therefore, the end marker related timers at the source target eNB expire and all handover processes are terminated accordingly.

Intra-eNB HO

The intra-eNB data handling is described in chapter 2.4.5.2. This section summarizes the implementation procedure within the LTE simulator. The intra-eNB handover is explained using the following example scenario where UE1 moves from cell 1 to cell 2 within the same eNB (eNB1). Figure B-22 shows that the UE is moving from cell1 to cell 2.

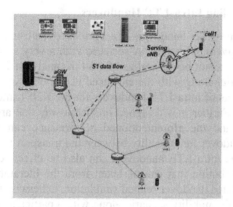

Figure B-22: Intra-eNB handover example scenario-1

The data to UE1 is received from an Internet node over the transport network. The eNB-1 serves the UE with data based on its channel capacity. When a handover indication is triggered, eNB-1 stops sending until the mobile node connects to cell 2.

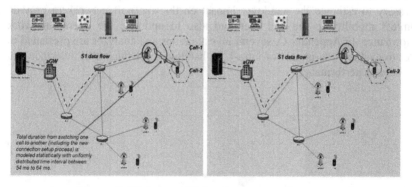

Figure B-23: Intra-eNB handover example scenario-2

This time duration is selected based on a statistical distribution. During this handover, the user buffer is reallocated based on the new cell number. Further, all parameters such as the data rate, are reset according to the new handover setting at cell 2. Figure B-23 illustrates the intra-eNB handover procedure graphically.

As in the inter-eNB handovers, mobility of all UEs is performed by the respective node in the mobility network model and provide the required handover information to the transport network through the central database.

Summary of Modeling Intra-LTE Handovers

In general, the intra-LTE handovers are correlated with a particular mobility model and user movement pattern. A scenario with a very specific mobility and propagation model cannot be generalized for all the analysis. Therefore, the intra-LTE handover model along with the enhanced user mobility model is developed to analyze the effects of intra-LTE handovers for the X2/S1 interface and the end user performance. Further it is designed in such a way that any mobility model can be deployed and the aforementioned performance can be evaluated. By decoupling the handover functionality between the transport network and the UE mobility model, the intra-LTE handovers can also be effectively modeled using any statistical distribution that can be taken from the literature or from a trace taken from a dedicated handover related simulation. Otherwise the UE mobility is restricted to a certain mobility scenario along with a particular movement pattern of the mobile nodes so that it does not provide the expected handover behaviors to the transport network. The average UE handover frequency is one of the key considerations that lead to additional bursty load to the transport network during handovers. Such a spontaneous bursty traffic can create congestion in the transport network resulting in a lot of packet losses if the appropriate measures have not been taken at the design stage of the model development. Therefore, it is necessary to estimate the handover impact on the transport network to provide seamless mobility to the end user and also to optimize the overall network performance. In Appendix-C, several investigations and analyzes are presented in order to evaluate the impact of intra-LTE handovers for the X2/S1 interface and the end user performance.

Appendix-C. LTE Transport Performance Analysis

The main objective of this investigation is to evaluate the impact of the LTE transport network (X2 and S1) on the end-to-end user performance. The LTE transport network plays a key role when the user mobility along with intra-eNB and inter-eNB handovers is considered in order to provide seamless mobility for the end users. Traffic differentiation based on transport bearer mapping and prioritization of X2 data forwarding has to be provided over the transport network to meet end user QoS requirements during handovers. Often, a congestion situation of the transport network causes severe impact on lower priority best effort traffic which is run on top of the TCP protocol. To the best of the author's knowledge, there is no literature which focuses on this area of investigation and analysis yet.

Three main applications, VoIP, web browsing and FTP downloading are used for the LTE transport investigation along with the inter-LTE handovers which include inter-eNB handovers and intra-eNB handovers. By applying the aforementioned traffic models, the impact of the LTE transport network on the end-to-end performance is investigated by categorizing the analysis into three main considerations.

- **Applying different transport bearer mapping:** the three different transport bearers, BE, AF1x and EF which have different priorities are configured at the LTE transport network and the end user performance is evaluated.

- **Applying different transport priorities for X2 forwarding data:** during inter-eNB handovers, buffered data at the source eNB which is referred to as the forwarded data has to be sent to target eNB promptly via the X2 interface in order to provide seamless mobility for the end user. From the source eNB point of view, forwarded data is sent in uplink direction whereas from the target eNB point of view, forwarded data is received in downlink direction. Since the uplink last mile is often in congestion, forwarded data may be delayed if priorities are not allocated. Therefore, effects of such situations are investigated by applying different transport priorities for the forwarded data.

- **Applying different transport congestion situations:** in the analysis, different transport congestion situations are applied by limiting the available capacity at the last mile network without changing the offered load. Depending on the level of congestion at the last mile, end user performance is investigated within the focus of this analysis.

All the aforementioned effects are analyzed together using the LTE system simulator which was described in Appendix-B. The section is organized as follows. Section C.1 presents the definition of traffic models which used for this analysis. The simulation parameter configuration and the simulation scenario definitions are described in section C.2. Later, the simulation results are presented discussing the effects on the end user performance in section C.3 and finally, a conclusion is given by summarizing all LTE based investigations and analysis.

C.1   Traffic Models

Three main traffic models are used for the LTE analysis. VoIP (Voice over IP) which is one of the popular real time applications is selected as the real time traffic model whereas the most extensively used web browsing (HTTP) and the FTP traffic models are selected as the best effort traffic models.

Real Time Traffic Model (Voip)

In this simulation analysis, one of the most commonly used voice codec, the GSM Enhanced Full Rate (EFR) is used. GSM EFR is one of the Adaptive Multi-Rate (AMR) codec family members with 12.2 kbps data rate at the application layer without considering the container overhead. GSM EFR coded frames are transported using the RTP/UDP protocol and encapsulated as one voice frame per IP packet. Further details of the used VoIP traffic can be summarized as follows.

- Codec: AMR 12.2 kbps (GSM EFR);
- No silence suppression;
- 50 frames/sec, 1 frame/IP packet;
- codec delay: 40 ms, Transport: RTP/UDP;
- Traffic idleness: 0.5 sec, Call duration: const. 90 sec;
- Call inter-arrival time: const. 90 sec and neg. exp. with 50 sec mean value.

Best Effort Traffic Models (HTTP, FTP)

The HTTP traffic model is deployed to emulate general web browsing applications which are vastly used in day to day applications whereas the FTP traffic model is used for large file downloads. These two traffic models are readily available in the default OPNET simulator environment, however they are modified and configured in such way that they are suitable for the focused analysis and investigations.

The HTTP traffic model is modified in such a way that the reading time does not include the statistics of the page download time and each web page is downloaded by a separate TCP connection. All the objects within a page use the already opened TCP connection which was used to download the main part of the page. The main parameter configuration for the HTTP traffic model is given below.

- Number of pages per session = 5 pages;
- Average page size = 100,000 bytes;
- Number of objects in a page = 1 first Object (frame) = 1 kbyte;
- 1 object = 100 kbyte;
- Reading time (default) = 12 sec (reading time does not include the page downloading time. The reading timer starts after completing the previous page).
- Each page uses a separate TCP connection and all objects within the page are downloaded via that single TCP connection.

The FTP traffic model was modified to achieve a maximum of three simultaneously active FTP downloads per UE; furthermore, each file download starts with a separate TCP connection. This means the UE can download three large files simultaneously from a server in the Internet using its broadband connection. The following parameters have been configured for the FTP traffic models:

- File size is 5Mbyte;
- The time duration between the start of a file download and the start of the next download is uniformly distributed between 30 and 60 sec.

C.2   Simulation Configuration

This section provides the details about the simulation setup and the main configuration parameters.

Simulation Setup and User Constellation

The 4 eNBs based reference simulation model which has been discussed in Appendix B.1 and Appendix B.2 is used with different user constellations. Mainly a 10 UEs / cell constellation (120 UEs / 4 eNBs) is used by deploying different transport configurations along with aforementioned traffic models. The LTE transport network which is the network between the aGW and the eNBs consists of several intermediate routers. The simulator overview is shown in Figure B-3. There are three IP based routers in the transport network of the current LTE simulator. The last mile is the link between an end-router which is R2 or R3 for this network and the eNBs. The last mile routers R2 and R3 are configured with DL transport PHBs which have been discussed in chapter 2.4 with limited transport capacities. Since the router network is not within the responsibility of the mobile network operators, the diffserv model which is configured in the router network cannot be controlled and therefore, the generic diffserv model implemented in the standard OPNET is used to configure the DL PHBs. However, the DL is also configured with the same diffserv parameters as in the UL diffserv model which has been developed in Appendix-B.

Parameter Configuration and Simulation Scenarios

For the user mobility, the enhanced random directional (RD) mobility model is used and the UEs are moving with a constant speed of 50 km/h. One UE is configured with one service type throughout the simulation period. The common simulator configuration consists of the following fixed parameter settings.

PDCP layer parameters (for UL and DL):

- Bearer buffer size = infinite.

Transport scheduler and shaper (only for UL):

- Scheduling algorithms = SP and WRED;
- Shaping rate (CIR) = 10 Mbit/sec;
- CBS = 1522 bytes;
- Input buffer size = 4000 bytes.

Link configurations:

- All links connected to the eNB are configured as Ethernet links with 10 Mbit/sec.
- All other links (mainly between routers) are configured with Ethernet links of 100 Mbit/sec.

General MAC layer parameters (for UL and DL):

- 6 schedulers per eNB (2 per cell for UL and DL).

For simplicity, the simulation scenarios are defined by dividing them into three categories which are listed below and discussed in the following.

- Transport configuration based on UE bearer mapping;
- X2 traffic prioritization configuration;
- Configuration based on different last mile capacities.

Transport Configuration Based on UE Bearer Mapping

The transport bearer mapping is done in the transport network of E-UTRAN for both UL and DL. As mentioned previously, the current traffic models are configured with one bearer per UE. Therefore one UE runs one type of service (application) at a time and the UE traffic is assigned to one DSCP value which shows the corresponding QCI configuration for that service. Mainly three types of transport bearers: EF, QCI 8 (AF1x) and QCI 9 (BE) are used for analyzes. There are four configurations named conf-1, conf-2, conf-3 and conf-4. The detailed transport configurations in which traffic flows are mapped into transport bearers are as follows:

Conf-1: All users are basic users and assigned to the default bearer with QCI = 9 (BE).

- VoIP users (basic) → QCI = 9 (BE),
- HTTP users (basic) → QCI = 9 (BE),
- FTP users (basic) → QCI = 9 (BE).

Conf-2: VoIP and HTTP users are assigned to the default bearer with QCI = 8 (AF11) and FTP users are assigned to the default bearer with QCI = 9 (BE).

- VoIP users (premium) → QCI= 8 (AF11),
- HTTP users (premium) → QCI= 8 (AF11),
- FTP users (basic) → QCI= 9 (BE).

Conf-3: VoIP users are assigned to the default bearer with QCI= 8 (AF11) and HTTP and FTP users are assigned to the default bearer with QCI= 9 (BE).

- VoIP users (premium) → QCI = 8 (AF11),
- HTTP users (basic) → QCI = 9 (BE),
- FTP users (basic) → QCI = 9 (BE).

Conf-4: VoIP users are assigned to the highest transport priority bearer with EF, HTTP users are assigned to the default bearer with QCI = 8 (AF11) and FTP users are assigned to the default bearer with QCI = 9 (BE).

- VoIP users (premium) → QCI = 1 (EF),
- HTTP users (basic) → QCI = 8 (AF1x),

- FTP users (basic) → QCI = 9 (BE).

Transport Prioritization of the X2 Data Forwarding

The HO performance is analyzed by applying different transport priorities at the X2 interfaces for the forwarding traffic. The X2 interface is a part of the LTE transport network which represents the transmission between eNBs. Whenever a UE performs a handover from one eNB (called source eNB) to another eNB (called target eNB), all PDCP data which is buffered at the source eNB has to be forwarded to the target eNB within a short period of time. Based on the transport priorities, the delay for forwarding data over the X2 interface varies. During these investigations, effects on the end user performance are analyzed by applying different priorities to the forwarded data. There are three cases which have been considered for this analysis.

a) All forwarded data has the *same* transport priority (called default priority) as the service priority.
b) All forwarded data has a *higher* transport priority as the service priority.
c) All BE (HTTP and FTP) forwarded data is *discarded* at the corresponding source eNB.

All above three options are configurable in the simulator for the corresponding investigations.

Different Last Mile Capacities

In order to create different congestion situations in the transport network, different last mile link capacities were configured for these simulations. For each case, the end user and the network performance were evaluated by setting up different simulation scenarios. Last mile capacities of 10 Mbit/sec, 8 Mbit/sec and 7 Mbit/sec were used for this analysis. All simulation scenarios which are defined based on above three categories are summarized in Table C-1.

Table C-1: LTE simulation scenarios

configurations	Last mile capacities		
	10 Mbps	8 Mbps	7 Mbps
conf-1	a, b, c	a, b, c	a, b, c
conf-2	a, b, c	a, b, c	a, b, c
conf-3	a, b, c	a, b, c	a, b, c
conf-4	a, b, c	a, b, c	a, b, c

The next section provides the simulation results for all scenarios given above.

C.3   Simulation Results Analysis

Since simulation analyzes and investigation provides a large number of statistics at different levels, this section summarizes and presents the simulation results by focusing on the end-to-end performance.

Voip End-to-End Delay in Sec

The average VoIP end-to-end delays for all users are shown in Figure C-1. The highest VoIP end-to-end performance is shown by the simulation scenarios with conf-4 in which the strict transport priority is allocated to the VoIP users.

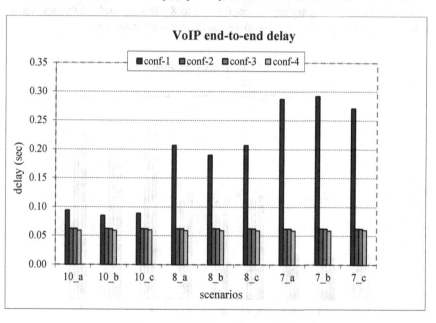

Figure C-1: VoIP end-to-end delay

It is shown that in order to achieve a satisfactory performance for VoIP, at least a sufficient BW should be configured or otherwise the connection should be mapped to a higher priority bearer at the transport network. All other scenarios which are based on conf-2, conf-3 and conf-4 show a good end-to-end performance for the VoIP traffic.

HTTP Page Download Time

Figure C-2 depicts the average HTTP response times over all users for the different simulation configurations. In these cases, conf-2 and conf-4 show the optimum end performance for the HTTP traffic. Even with a severe congestion situation (the last mile is configured to 7 Mbit/sec), HTTP users exhibit a very good end-to-end performance. This is due to the fact that HTTP traffic has higher priority over the transport network compared to the FTP traffic which normally downloads large files. The results of the other two configurations which are configured with equal transport priorities for FTP and HTTP traffic at the transport network further justify above conclusion. The congested scenarios in which the last mile capacities are configured to 8 Mbit/sec and 7 Mbit/sec show relatively high response times compared to the non-congested cases.

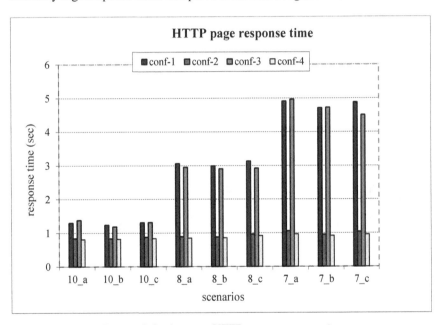

Figure C-2: Average HTTP page response time

DL FTP Downlink Time

The average FTP download times over all users are shown in Figure C-3. As expected, the figure shows that the download time increases when the last mile capacity is reduced. The last mile capacity of the 8 Mbit/sec based simulation scenario shows a higher download time compared to the 10 Mbit/sec based scenario, however in relation to the bandwidth reduction, the 7 Mbit/sec based simulation scenario depicts a significantly higher download time compared to other scenarios. This means, beyond a certain limit of the last mile capacity, for example 7 Mbit/sec in comparison to the 8 Mbit/sec, the FTP performance is severely degraded due large packet losses at the transport network.

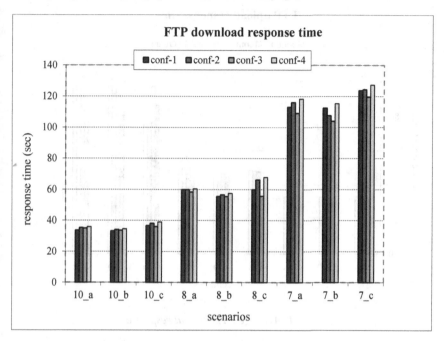

Figure C-3: Average FTP downloads response time

When granting priority to HTTP traffic at the transport network, conf-2 and conf-4 have no impact on the FTP end user performance. This is mainly due to the fact that the FTP users download very large files so that small delays of one or several retransmitted TCP packets do not significantly affect the total file download time. Further, the outcome of these simulation results show that the three different X2 forwarding options (a, b and c) do not have a significant impact on the end users performance.

UL FTP Upload Response Time

Figure C-4 shows the average FTP upload time over all users. The results are similar to the previous FTP download time. The highest upload response time is depicted by the scenarios with a congested last mile capacity of 7 Mbit/sec for all configurations. There is no significant difference of the FTP end user performance by applying different X2 forwarding priorities during UE HOs. As discussed above, the FTP end user performance degrades when limiting the last mile capacity.

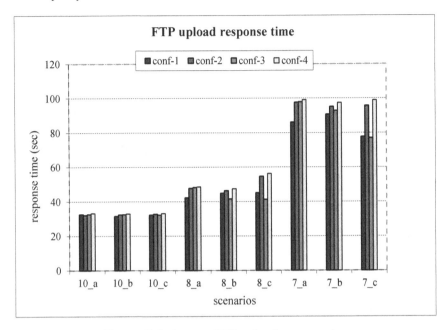

Figure C-4: Average FTP upload response time

DL Total Number of Packet Losses

Figure C-5 shows the total number of packet losses for all users. These statistics include all packet losses due to the buffer overflow and due to HOs such as RLC packet losses in the eNB entity. Figure C-5 clearly depicts that the 7 Mbit/sec last mile capacity shows the highest number of packet losses which is mainly due to the congestion at the transport network.

The highest number of packet discards is shown in the simulation scenarios which are configured with the X2 forwarding option "c" for any last mile capacity configuration whereas the lowest number of packet discards is shown in case of forwarding option "b" since the HTTP traffic is granted a high priority at the transport network.

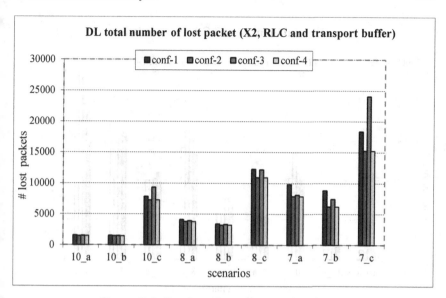

Figure C-5: Total number of lost packets for DL

Total Number of TCP Retransmissions

The total number of TCP retransmissions for the DL is shown in Figure C-6. The TCP retransmissions are either due to TCP time-outs or due to lower layer packet losses. The configuration of 7 Mbit/sec last mile capacity causes severe congestion at the transport network which results in a large number of packet losses and consequently a large number of TCP retransmissions. Each TCP retransmission always results in adding more overhead to the transport network and reduces the achievable goodput.

A lower number of TCP retransmissions can be observed for X2 forwarding options "a" and "b" compared to option "c" for all last mile capacities. This is mainly due to a higher number of packet discards by activating X2 forwarding option "c" (discard timer is activated) during HOs. All other scenarios depict a relatively low number of TCP retransmissions depending on the degree of congestion level at the last mile of the transport network. A last mile capacity of 7 Mbit/sec shows the highest number of TCP retransmissions which corresponds to the number of packet discards at the transport network due to congestion.

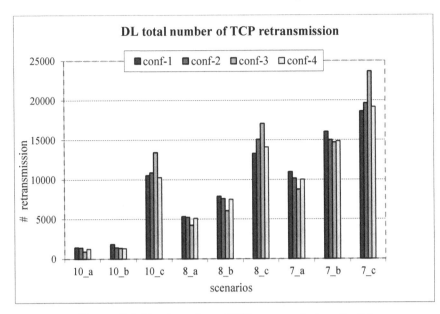

Figure C-6: Total number of TCP retransmissions for DL

Throughput between R1 and aGW

Figure C-7 shows the overall average DL throughputs between aGW and the R1 router. The throughput statistic includes all carried load of the last mile transport networks including the upper layer retransmissions and overheads. Further it shows the statistical multiplexing gain of the carried load for the different last mile capacities. The simulation of the highly congested transport link with 7 Mbit/sec last mile capacity depicts a comparatively slightly lower load carried over the transport network.

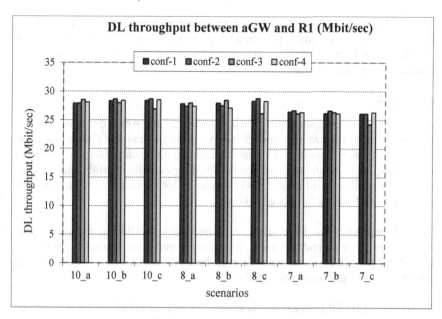

Figure C-7: DL link throughput between aGW and R1

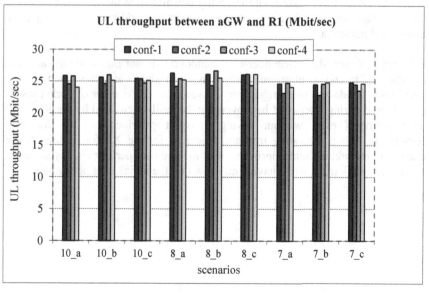

Figure C-8: UL link throughput between R1 and aGW

The average UL link throughput between the R1 router and aGW is shown Figure C-8 which depicts a similar performance as in the DL throughput. There is no significant difference of the carried load among the simulation scenarios compared to the DL results.

Conclusion

A comprehensive LTE simulator which was discussed in Appendix-B was used to investigate and analyze the end-to-end performance under different transport effects such as transport level bearer mapping, QoS classification, transport congestion, LTE handovers and bandwidth dimensioning. Mainly three traffic types were used with different QoS assignments for simulation analyzes and the following key findings have been achieved.

To guarantee QoS for VoIP users, it is necessary to allocate separate bearer mapping at the transport network. Whenever a particular transport priority or bearer mapping is assigned together with best effort traffic, the VoIP end user performance is significantly reduced.

The optimum HTTP performance is achieved when the HTTP traffic is separated and given a higher priority than large file downloads at the transport network. When HTTP traffic is mixed with large file downloads in the transport network, the web page response time is significantly increased and provides a very poor end-to-end performance.

Simulation results show that inter-eNB handovers do not play a significant role on the end-to-end application performance. In these analyzes each UE is performing handovers in average every 45 seconds. The number of packet losses experienced during inter-eNB handovers can be easily recovered by the RLC and TCP protocols even without having transport priorities for the X2 data forwarding. When allocating transport priorities for the X2 forwarding in the transport network, a slight improvement of the end user performance is achieved for the downlink users while having no impact on the uplink user performance.

References

[1] T. Weerawardane, A. Timm-Giel, C. Görg, T. Reim, "Performance Analysis of the Iub Interface in UMTS Networks for HSDPA", Mobilfunk-Technologien und Anwendungen, Vorträge der 10. ITG-Fachtagung, Osnabrück, Juni 2005.

[2] L. Zhao, T. Weerawardane, A. Timm-Giel, C. Görg, U. Türke, M. Koonert "overview on UMTS HSDPA and Enhanced Uplink (HSUPA)", Mobilfunk-Technologien und Anwendungen, Vorträge der 11. ITG-Fachtagung, Osnabrück, May 2006.

[3] T. Weerawardane, A. Timm-Giel, C. Görg, T. Reim, "UMTS Transport Network Layer: Modeling and Optimization of HSDPA Traffic Flows", IEE conference, Sri Lanka, Sep. 2006.

[4] T. Weerawardane, A. Timm-Giel, C. Görg, T. Reim, "Impact of the Transport Network Layer Flow Control for HSDPA Performance", IEE conference, Sri Lanka, Sep. 2006.

[5] T. Weerawardane, X. Li, A. Timm-Giel, C. Görg, "Modeling and Simulation of UMTS HSDPA in OPNET", in Proc. OPNETWORK, Washington DC, USA, September 2006.

[6] M. Becker, T. Weerawardane, X. Li, C. Görg, "Recent Advances in Modelling and Simulation Tools for Communication Networks and Services, Extending OPNET Modeller with External Pseudo Random Number Generators and Statistical Evaluation by the Limited Relative Error Algorithm", chapter 12, pages 241-256, 1st edition, 2007.

[7] T. Weerawardane, A. Timm-Giel, G. Malafronte, G. Durastante, Stephan Hauth, C. Görg, "Preventive and Reactive based TNL Congestion Control Impact on the HSDPA Performance", VTC IEEE conference, Singapore, May 2008.

[8] T. Weerawardane, A. Timm-Giel, G. Malafronte, G. Durastante, S. Hauth, C. Görg, "Effective Iub Congestion Control for the HSDPA Performance", ICT Mobile Summit, Stockholm, Sweden, June 2008.

[9] T. Weerawardane, R. Perera, A. Timm-Giel, C. Görg, "A Markovian Model for HSDPA TNL Congestion Control Performance Analysis", VTC IEEE conference, Canada, Sept. 2008.

[10] X. Li, Y. Zaki, T. Weerawardane, A. Timm-Giel, C. Görg, "HSUPA Backhaul Bandwidth Dimensioning", in 19th IEEE International Symposium on Personal, Indoor and Mobile Radio Communications (PIMRC), Cannes, France, Sept. 2008.

[11] Y. Zaki, T. Weerawardane, X. Li, A. Timm-Giel, G. Malafronte, C. Görg, Effect of the RLC and TNL Congestion Control on the HSUPA Network Performance, Mishawaka International Conference on Communications, Computers and Applications, MIC-CCA in Amman, Jordan, Jan. 2008.

[12] T. Weerawardane, Y, Zaki, A. Timm-Giel, G. Malafronte, S. Hauth, C. Görg., "Effect of the TNL congestion control for the HSUPA performance", IEEE International VTC Spring in Barcelona, Spain, April 2009.

[13] T. Weerawardane, A. Timm-Giel, C. Görg, G. Malafronte, S. Hauth, "Effect of TNL flow control schemes for the HSDPA network performance", special issues-2 of Journal of Communications (IEEE), Academy Publisher, Finland, pages 78-88, 2009.

[14] X. Li, Y. Zaki, T. Weerawardane, A. Timm-Giel, C. Görg, G. Malafronte, "Use of Traffic Separation Techniques for the Transport of HSPA and R99 Traffic in the Radio Access Network with Differentiated Quality of Service", IJBDCN Journal, pages 84-100, July–Sept., 2009.

[15] T. Weerawardane, A. Timm-Giel, C. Görg, G. Malafronte, S. Hauth, "Effect of TNL Congestion Control Schemes on the HSDPA Network Performance" Journal Publication, IEEE VTC magazine, Vehicular Technology Magazine, IEEE Volume 4, pages 54-63, December 2009.

[16] Y. Zaki, T. Weerawardane, A. Timm-Giel, C. Görg, G. Malafronte, "Performance Enhancement due to the TNL Congestion Control on the Simultaneous Deployment of both HSDPA and HSUPA", Special Issue: Recent Advances in Communications and Networking Technologies, JNW Journal, Vol 5, No 7 (2010), pages 773-781.

[17] H. Holma, A. Toskala, "WCDMA for UMTS, John Wiley & Sons", Inc., Chichester, UK, 3rd edition, 2004.

[18] H. Holma, A. Toskala, "HSDPA/HSUPA for UMTS: High Speed Radio Access for Mobile Communications", John Wiley & Sons, Chichester, UK, 2006.

[19] ETSI, Universal Mobile Telecommunications System (UMTS): Selection procedures for the choice of radio transmission technologies of the UMTS (UMTS 30.03 version 3.2.0), TR 101 112 v3.2.0, Apr. 1998.

[20] F. Agharebparast, V. Leung, "QoS support in the UMTS/GPRS backbone network using DiffServ", IEEE Global Telecommunications Conference, vol. 2, pages 1440–1444, Nov. 2002.

[21] R. Fielding, J. Gettys, J. Mogul, H. Frystyk, L. Masinter, P. Leach, T. Berners-Lee, "Hypertext Transfer Protocol - HTTP/1.1", RFC 2616, Internet Engineering Task Force, June 1999.

[22] 3GPP TS 25.855 High Speed Downlink Packet Access (HSDPA): Overall UTRAN description, 3GPP TS 25.856 HSDPA: Layer 2 and 3 aspects, 3GPP TS 25.876 Multiple-Input Multiple-Output Antenna Processing for HSDPA, 3GPP TS 25.877 HSDPA-Iub/Iur Protocol Aspects, 3GPP TS 25.890 HSDPA: User Equipment (UE) radio.

[23] ETSI, Universal Mobile Telecommunications System (UMTS): Physical layer aspects of UTRA High Speed Downlink Packet. 3GPP, TR 25.848 v3.2.0, 2001-03.

[24] M. Necker, A. Weber, "Impact of Iub Flow Control on HSDPA System Performance", proceedings of the 16th Annual IEEE International Symposium on Personal Indoor and Mobile Radio Communications, Berlin, Germany 2005.

[25] M. Necker, A. Weber, "Parameter selection for HSDPA Iub flow control", in Proc. 2nd International Symposium on Wireless Communication Systems (ISWCS 2005), Siena, Italy, September 2005.

[26] G. Aniba, S. Aissa, "Adaptive proportional fairness for packet scheduling in HSDPA", in Global Telecommunications Conference, GLOBECOM 2004, vol. 6, December 2004, pages 4033-4037.

[27] P. Legg, "Optimized Iub flow control for UMTS HSDPA", In Proc. IEEE Vehicular Technology Conference (VTC 2005-Spring), Stockholm, Sweden, June 2005.

[28] R. Comroe, D. Costello, Jr., "ARQ schemes for data transmission in mobile radio systems", IEEE Journal on Selected Areas in Communications, vol. 2, no. 4, pages 472-481, July 1984.

[29] S. Floyd, M. Handley, J. Padhye, "A Comparison of Equation-Based and AIMD Congestion Control", ACIRI May 12, 2000.

[30] V. Jacobson, M. Karels, "Congestion Avoidance and Control", University of California at Berkeley, USA, November, 1988.

[31] A. Klemm, C. Lindemann, M. Lohmann, "Traffic Modeling and Characterization for UMTS Networks", Proc. IEEE Globecom 2001, San Antonio Texas, USA, November 2001.

[32] X. Li, R. Schelb, C. Görg, A. Timm-Giel, "Dimensioning of UTRAN Iub Links for Elastic Internet Traffic", International Teletraffic Congress, Beijing, Aug/Sept., 2005.

[33] X. Li, R. Schelb, C. Görg, A. Timm-Giel, "Dimensioning of UTRAN Iub Links for Elastic Internet Traffic with Multiple Radio Bearers", in Proc. 13th GI/ITG Conference Measuring, Modeling and Evaluation of Computer and Communication Systems, Nürnberg, March 2006.

[34] X. Li, S. Li, C. Görg, A. Timm-Giel, "Traffic Modeling and Characterization for UTRAN", in Proc. 4th International Conference on Wired/Wireless Internet Communications, Bern Switzerland, May 2006.

[35] X. Li, R. Schelb, A. Timm-Giel, C. Görg, "Delay in UMTS Radio Access Networks: Analytical Study and Validation", in Proc. Australian Telecommunication Networks and applications conference (ATNAC), Melbourne Australia, December 2006.

[36] X. Li, L. Wang, R. Schelb, T. Winter, A. Timm-Giel, C. Görg, "Optimization of Bit Rate Adaptation in UMTS Radio Access Network", in Proc. 2007 IEEE 65th Vehicular Technology Conference VTC2007-Spring, Dublin, Ireland, April 2007.

[37] X. Li, W. Cheng, A. Timm-Giel, C. Görg, "Modeling IP-based UTRAN for UMTS in OPNET", distinguished paper award, in Proc. OPNETWORK 2007, Washington DC, USA, September 2007.

[38] X. Li, Y. Zeng, B. Kracker, R. Schelb, C. Görg, A. Timm-Giel, "Carrier Ethernet for Transport in UMTS Radio Access Network: Ethernet Backhaul Evolution", 2008 IEEE 67th Vehicular Technology Conference VTC2008-Spring, Singapore, May 2008.

[39] 3GPP technical specification group services and system aspects, "Evolved Universal Terrestrial Radio Access (E-UTRA) and Evolved Universal Terrestrial Radio Access Network", 3GPP TS 36.300 V8.5.0 (2008-05): Overall description, 3GPP TS 36.300 V8.5.0 (Rel'8), June 2008.

[40] 3GPP Technical Specification Group Radio Access Network "Requirements for Evolved UTRA (E-UTRA) and Evolved UTRAN (E-UTRAN)", 3GPP TR 25.913 V9.0.0 (2009-12)-release 9.

[41] R. Stewart, "Stream Control Transmission Protocol", RFC 4960, September 2007.

[42] P. Lescuyer, T. Lucidarme, "Evolved Packet System (EPS): The LTE and SAE Evolution of 3G UMTS", John Wiley & Sons, Ltd., 2008.

[43] E. Dahlman, S. Parkvall, J. Sköld, P. Beming, "3G Evolution HSPA and LTE for Mobile Broadband", Academic Press of Elsevier,1st edition, 2007.

[44] H. Holma, A. Toskala, "WCDMA for UMTS-HSPA Evolution and LTE" 4th edition, John Wiley & Sons, Ltd., 2007.

[45] S. Sesia, I. Toufik, M. Baker, "The UMTS Long Term Evolution from Theory to Practice", 1st edition, John Wiley & Sons Ltd., 2009.

[46] 3GPP technical specification group services and system aspects, "General Packet Radio Service (GPRS) enhancements for Evolved Universal Terrestrial Radio Access Network (E-UTRAN) access", 3GPP TS 23.401 V8.2.0 (Rel'8), June 2008.

[47] 3GPP Technical Specification Group Radio Access Network, "Evolved Universal Terrestrial Radio Access (E-UTRA) and Evolved Universal Terrestrial Radio Access Network (E-UTRAN): Radio Interface Protocol Aspects", 3GPP TR 25.813 v7.1.0, 2006.

[48] 3GPP technical specification group services and system aspects, "Evolved Universal Terrestrial Radio Access (E-UTRA) Medium Access Control (MAC) protocol specification", 3GPP TS 36.321 V8.2.0 (Rel'8), May 2008.

[49] 3GPP technical specification group services and system aspects, "Evolved Universal Terrestrial Radio Access (E-UTRA); Packet Data Convergence Protocol (PDCP) specification", 3GPP TS 36.323 V8.2.1 (Rel'8), May 2008.

[50] MEF Technical specification 10.1 and MEF Technical specification 10.3, Online-web: http://metroethernetforum.org/index.php.

[51] T. Camp, J. Boleng, V. Davies, "A Survey of Mobility Models for Ad-Hoc Network Research", Wireless Communications & Mobile Computing (WCMC): Special issue on Mobile Ad Hoc Networking: Research, Trends and Applications, Vol. 2, No. 5. (2002), pages 483-502.

[52] T. Gayathri, S. Venkadajothi, S. Kalaivani, C. Divya, C. Dhas, "Mobility management in next-generation wireless systems", MASAUM Journal of Computing, volume-1, issue-2, September 2009.

[53] J. Markoulidakis , G. Lyberopoulos, D. Tsirkas , E. Sykas, "Mobility Modelling in Third-Generation Mobile Telecommunications Systems", IEEE Personal Communications, Vol. 4, No. 4, pages 41-56, 1997.

[54] E. Dahlman, S. Parkvall, J. Sköld, P. Beming, 3G Evolution: HSPA and LTE for mobile broadband, first edition, 2007.

[55] 3GPP, Technical Specification Group RAN, Synchronization in UTRAN Stage 2 (Release 1999), 3GPP TS 25.402 version 3.10.0, June. 2002.

[56] 3GPP, Technical Specification Group RAN, UTRAN overall description 3GPP TS 25.401 version 3.10.0, June 2002.

[57] 3GPP, Technical Specification Group RAN, Packet Data Convergence Protocol (PDCP) specification, 3GPP TS 25.323 version 3.10.0, Sep. 2002.

[58] 3GPP, Technical Specification Group RAN, Radio Link Control (RLC) protocol specification, 3GPP TS 25.322 version 3.18.0, June 2004.

[59] 3GPP, Technical Specification Group RAN, Medium Access Control (MAC) protocol specification, 3GPP TS 25.321 version 3.17.0, June 2004.

[60] 3GPP, Technical Specification Group RAN, "Feasibility study for Enhanced Uplink for UTRA FDD", 3GPP TR 25.896 version 6.0.0, March 2004.

[61] 3GPP, Technical Specification Group RAN, FDD Enhanced Uplink; Overall Description; Stage 2, 3GPP TS 25.309 version 6.6.0, Apr. 2006.

[62] ITU-T Recommendation Q.2931.1: ITU-T Recommendation I.363.1 (1996), B-ISDN ATM Adaptation Layer specification: Type 1 AAL. ITU-T Recommendation I.363.2 (1997), B-ISDN ATM Adaptation Layer specification: Type 2 AAL. ITU-T Recommendation I.363.3 (1996), B-ISDN ATM Adaptation Layer specification: Type 3/4 AAL.ITU-T Recommendation I.363.5 (1996), B-ISDN ATM Adaptation Layer specification: Type 5 AAL.

[63] J. Beckers, I. Hendrawan, R. Kooij, R. van der Mei, "Generalized Processor Sharing Performance Models for Internet Access Lines", In Proc. of 9th IFIP Conference on Performance Modeling and Evaluation of ATM & IP Networks, Budapest, Hungary, 2001.

[64] ITU-T Recommendation G.992.1 (1999): Asymmetric digital subscriber line (ADSL) transceivers, G.992.2 (1999): Splitterless asymmetric digital subscriber line (ADSL) transceivers, G.992.2 (2002): Asymmetric digital subscriber line transceivers 2 (ADSL2), G.992.2 (2003):Asymmetric Digital Subscriber Line (ADSL) transceivers - Extended bandwidth ADSL2 (ADSL2plus).

[65] ITU-T Recommendation G.992.2 (1998): High bit rate Digital Subscriber Line (HDSL) transceivers.

[66] R. Atkinson, S. Kent, "Security Architecture for the Internet Protocol", RFC 2401, April 2010.

[67] OPNET Modeler, available at http://www.opnet.com, accessed on April 2009.

[68] W. Simpson, "The Point-to-Point Protocol (PPP)", RFC 1661, Nov. 2009.

[69] W. Simpson, "PPP in HDLC-like Framing", RFC 1662 Nov. 2009.

[70] J. Lau, M. Townsley, "Layer Two Tunneling Protocol version-3 (L2TPv3)", RFC 3931, Nov. 2009.

[71] W. Townsley, A. Valencia, "Layer Two Tunneling Protocol (L2TP)", RFC 2661, Dec. 2009.

[72] L. Mamakos, K. Lidl, J. Evarts, A. Valencia, "A Method for Transmitting PPP Over Ethernet (PPPoE)", RFC 2516, Dec. 2009.

[73] 3GPP technical specification group services and system aspects, "Evolved Universal Terrestrial Radio Access (E-UTRA): Radio Link Control (RLC) protocol specification", 3GPP TS 36.322 V8.2.0 (Rel'8), May 2008.

[74] B. Braden, S. Floyd, V. Jacobson, G. Minshall, S. Shenker, "Queue Management and Congestion Avoidance in the Internet", RFC 2309, April 1998.

[75] M. Christiansen, K. Jeffay, D. Ott, F.D. Smith, Tuning RED for Web Traffic, ACM SIGCOMM, August 2000/June 2001 version in IEEE/ACM Transactions on Networking.

[76] L. Le, J. Aikat, K. Jeffay, F.D. Smith, "The Effects of Active Queue Management on Web Performance", ACM SIGCOMM, 2003.

[77] J. Babiarz, K. Chan, "Configuration Guidelines for DiffServ Service Classes", RFC 4594, Nov. 2009.

[78] H. Wadsworth, "The Handbook of Statistical Methods for Engineers and Scientists", 1st edition, 1997.

[79] Communication Network Class Library, CNCL, University of Bremen, available at http://www.comnets.uni-bremen.de/docs/cncl/, accessed on May. 2010.

[80] J. Dongarra, P. Raghavan, "A new recursive Implementation of sparse Cholesky factorization", in Proceedings of the 16th IMACS World Congress 2000 on Scientific Computing, Applications. Mathematics, and Simulation, Lausanne, Switzerland, August 2000.